T0205970

Lecture Notes in Mathematics

Volume 2277

This series reports on new developments in all areas of mathematics and their applications - quickly, informally and at a high level. Mathematical texts analysing new developments in modelling and numerical simulation are welcome. The type of material considered for publication includes:

1. Research monographs
2. Lectures on a new field or presentations of a new angle in a classical field
3. Summer schools and intensive courses on topics of current research.

Texts which are out of print but still in demand may also be considered if they fall within these categories. The timeliness of a manuscript is sometimes more important than its form, which may be preliminary or tentative.

More information about this series at http://www.springer.com/series/304

Adriano M. Garsia • Ömer Eğecioğlu

Lectures in Algebraic Combinatorics

Young's Construction, Seminormal
Representations, $\mathfrak{sl}(2)$ Representations,
Heaps, Basics on Finite Fields

 Springer

Adriano M. Garsia
Department of Mathematics
University of California, San Diego
La Jolla, CA, USA

Ömer Eğecioğlu
Department of Computer Science
University of California
Santa Barbara, CA, USA

ISSN 0075-8434   ISSN 1617-9692   (electronic)
Lecture Notes in Mathematics
ISBN 978-3-030-58372-9   ISBN 978-3-030-58373-6   (eBook)
https://doi.org/10.1007/978-3-030-58373-6

Mathematics Subject Classification: 05E10, 20C30, 05A99, 11T06, 11T22, 11T30

This Springer imprint is published by the registered company Springer Nature Switzerland AG.
The registered company address is: Gewerbestrasse 11, 6330 Cham, Switzerland

Preface

When I went to the University of Rome in 1948 and heard for the first time my mathematics professors' presentations, I was totally fascinated by that subject. I still am, seven decades later.

I wanted to learn all the branches of this amazing subject. In particular, I wanted to know algebra which was not offered the year I registered. So I started to read the elementary book of Ugo Amaldi as an introduction to that branch. To my surprise, every time I read for more than 2 h that particular book I ended up with an intense headache... The surprise was the fact that many other books that I was also reading did not give me a headache no matter how many hours I spent on them.

I was anxious to know what caused this undeniable difference. One thing was quite obvious by comparing the class notes by Mauro Picone who was my Calculus professor with Amaldi's book: the latter contained mostly words and the former contained mostly formulas. Could it be that my brain was better disposed at grasping the content of formulas than parsing the meaning of sentences?

So I began to read the algebra literature. To my further surprise, I noticed that earlier algebra literature contained more formulas and later ones contained more words. Basically, the pre-Galois literature such as *Lagrange* contained more formulas and the post-Galois literature such as *Dedekind* contained more words... The conclusion converged to the realization that my mathematical talents were mostly visual. Words and sentences required more mnemonic capabilities that my brain apparently did not possess. That shaped my mathematical future ever after.

I found out when I immigrated to America how lucky I was to have ended up in Stanford. The distinguished Stanford professors of that time were Pólya, Szegö, Stephan Bergman, Loewner, Schiffer, and the promising younger Garabedian. There was a variety of classes offered in Complex Variables, Classical Analysis, Partial Differential Equations but no Abstract Algebra and Topology.

Schiffer gave a course in Galois theory essentially reading the work of Lagrange and Galois' letter. I really enjoyed that course to the point that in my UCSD times I gave a similar course and I wrote lecture notes in Galoisian Galois theory. I learned later from Berlekamp that the difference between Galoisian Galois Theory and Dedekind's version of the latter was actually quite simple. From the abstract viewpoint, all Galois fields of the same dimension are isomorphic, but from the engineering point of view some were much more expensive than others... In fact, apparently some of them can be realized as a piece of hardware for a few thousand dollars and others may require as much as a thousand times more.

These early findings shaped my mathematics for the rest of my years. I concluded that mathematicians have different talents, depending on their genetic constructs. Thus, I chose my Ph.D. students according to their ability to facilitate their interactions with me. All my lecture notes emphasize that visual aspect in the abundance of formulas and the many illustrations that are included.

It was only later in formulating the definition of combinatorial structure that I really understood the nature of my mathematical abilities. I came to define a *Combinatorial Structure* as a visual realization of a *Mathematical Construct*. People with *visual talents* may excel in finding a variety of statistics of combinatorial structures that may not appear as obvious from the original Mathematical Construct.

An example should better get across this point. If we search for *Parking Function* in the internet we get, plus or minus, the following information. A PF is a sequence $a_1, a_2, \ldots, a_n$ with $1 \leq a_i \leq n$ whose weakly increasing rearrangement $b_1, b_2, \ldots, b_n$ has the property that $b_i \leq i$ for all $1 \leq i \leq n$. In all my writings, I depict a PF as a Dyck path in the $n \times n$ lattice square with the labels $1, 2, \ldots, n$ filling the cells adjacent to the North steps so that they increase along the vertical steps of the path. Over the years that I have been using this image, many talented combinatorists constructed the *area*, the *bounce*, the *dinv*, and many other features of a PF, none of which is visualizable from the original construct.

I would like to say a few words and set the stage for each of the lecture notes collected here.

Lecture 1: Alfred Young's Construction of the Irreducible Representations of S_n

Let us recall that the simplest *algebra* is the collection of $n \times n$ matrices with complex entries. By taking the collection of block diagonal matrices of sizes $n_1, n_2, \ldots, n_k$, we get what is usually called a *semisimple algebra*. Alfred Young, unaware of Frobenius' work, gave a completely explicit construction, within the algebra of S_n, of a family of elements indexed by a partition of n

$$\mathcal{E}^\lambda = \{e_{ij}^\lambda\}_{i,j=1}^{n_\lambda}$$

with multiplication table

$$e_{ij}^\lambda e_{rs}^\mu = \begin{cases} e_{is}^\lambda & \text{if } \lambda = \mu \text{ and } j = r \\ 0 & \text{otherwise} \end{cases}$$

and showed that the collection $\bigcup_{\lambda \vdash n} \mathcal{E}^\lambda$ is a basis of the group algebra $\mathcal{A}(S_n)$. In particular, he had to show the identity

$$n! = \sum_{\lambda \vdash n} n_\lambda^2 .$$

What is remarkable is that Young's construction was purely combinatorial and relatively simple. Young in doing this had proved the semisimplicity of $\mathcal{A}(S_n)$.

Of course, Frobenius had actually proved the same result for all finite groups. In particular, he had shown that k (the number of blocks) gives, for every finite group G, the dimension of the center of $\mathcal{A}(G)$. In the case $G = S_n$, Frobenius went further and defined what is now called the *Frobenius map F* that sends the center of $\mathcal{A}(S_n)$ onto the vector space $\Lambda^{=n}$ of symmetric functions homogeneous of degree n. This permitted Frobenius to identify the irreducible characters of S_n as $\chi^\lambda = F^{-1} s_\lambda$. In spite of the ambiguity in the case of self-conjugate partitions, Young's and Frobenius' formulas are the same. That is,

$$\frac{n!}{f_\lambda} \sum_{i=1}^{n_\lambda} e_{i,i}^\lambda = \chi^\lambda = F^{-1} s_\lambda .$$

The map F creates a bridge from the center of $\mathcal{A}(S_n)$ to the combinatorics of change of bases of $\Lambda^{=n}$. This circumstance opens up an immense variety of combinatorial problems that are at the heart of the power of Algebraic Combinatorics. This power is entirely overlooked by most researchers outside of the latter branch of Mathematics.

The reason is quite simple and became obvious at my obituary of Jeff Remmel, my colleague at UCSD in Algebraic Combinatorics who passed away prematurely. I lamented that in all the 50 years of our mathematical activity, we were never able to impress some of our colleagues by our results. They judged the size of our branch by what they knew about it. So they concluded that we worked on *next to nothing*.

The other reason is based on the widely accepted notion that the only thing that is worth proving is what is valid for all Coxeter groups. That is a fallacy due to a lack of understanding the deep properties of S_n and its Frobenius map. Alfred Young must have been quite aware of it when he extended his construction of his *matrix units* to B_n and A_n.

The Frobenius results show that matrix units do exist in $\mathcal{A}(G)$ of every finite group G. This should be possible for $G = GL[n]$ with entries in a finite field. However, the only result in this direction was proved by Macdonald in his identification of the center of $\mathcal{A}(G)$. This chapter of the lecture notes is unique since it is a simplified derivation of all of Young's work for S_n and ends with properties of the Frobenius map.

Lecture 2: Young's Seminormal Representation, Murphy Elements, and Content Evaluations

I do not understand now why I did not publish these lecture notes at the time I wrote them. It is a natural chapter to follow an exposition of Young's work since it uses the same operators discovered by Young. My presentation is obtained by following Young and later writings of Thrall, Rutherford, Jucys, and Murphy. In the text, I specifically show how the contents are related to the work of Macdonald, Goupil et al., and Diaconis-Greene.

The most surprising recent development is in a communication by Persi Diaconis who was highly complimentary of this work and was unaware of its existence, never mind the solutions to problems formulated in his joint work with Curtis Greene.

Following my exchange with Diaconis, I suspected that also Curtis Greene was not aware of the existence of this paper. So I sent him also a copy. The e-mail I received from him was gratifying as well, especially his realization that his paper with Persi had a significant follow-up in the applications of the Young Seminormal Units. Given the present situation, he mentioned that he will dedicate all his free time for a careful reading of my paper. I expect some further significant discoveries from Curtis after these readings. By looking at the introduction, one can certainly get a precise idea of the contents of this chapter. I worked very hard to make it clear and easy to read. This is a beautiful application of Algebraic Combinatorics at its best.

Lecture 3: On Finite Dimensional $\mathfrak{sl}(2)$ Representations and an Application to Algebraic Combinatorics

I wrote these lecture notes after I learned that Richard Stanley had proved the Spernerity of the $L[m, n]$ posets. I also learned that Robert Proctor had actually shown that Stanley's result could be proved using $GSL[2]$ theory. Since I knew very little about that branch of representation theory, I was very anxious to learn more about it to be able to understand the Proctor proof of Stanley's result.

Unfortunately, I quickly realized that the $GSL[2]$ results were scattered through a voluminous collection of Lie algebra results. Yet the little understanding that I had of this branch of mathematics convinced me that $GSL[2]$ theory should be totally developed using elementary linear algebra. To confirm this belief, I consulted my UCSD colleague Nolan Wallach for advice and guidance. I knew that Nolan was an expert in representation theory and I was eager to learn from him all that was known about $GSL[2]$ theory.

We quickly came to formulate a manner of constructing this elementary treatment. Nolan was to state a theorem and I tried to prove it using only elementary linear algebra. If I succeeded, I will write up my proof. If I did not succeed, he would then guide me into finding a way to do it. To make a long story short, the number of times he needed to help me makes it quite evident that these lecture notes would not come to exist without Nolan's help. But nevertheless, they would not come to be were it not also for my determination to write up a development of this branch of representation theory, presented in a manner that every mathematician can read it in full detail without excessive effort.

This done, I was ready to understand Proctor's proof of Stanley's Theorem. To complete my $GSL[2]$ lecture notes, in a manner that would be especially attractive to practitioners of Algebraic Combinatorics, there is an added last section with a proof of the Spernerity of the $L[m, n]$ posets.

Lecture 4: Heaps, Continued Fractions, and Orthogonal Polynomials

There is an interesting history related to these lecture notes. After studying the proof of the MacMahon Master Theorem given in the Cartier–Foata book, I decided to make their proof more visual by replacing their language approach by combinatorial objects. The letters in their language are the $n \times n$ matrix elements $b_{i,j}$ for $1 \leq i < j \leq n$. They state that the pair $b_{i,j}$ and $b_{i',j'}$ commute if and only if $i \neq i'$. To make their proof more visual, I expressed their commutativity statement by an object impossibility. To get across this step, we will let $n = 3$ and draw the noncommutativity of pairs that start with $i = 1$ by the following picture.

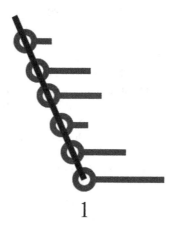

1

Rods of different lengths emanating from one cannot commute since they are threaded to a bar. To make a long story short, I translated their proof and changed it into a visual argument. This done, I showed the resulting imagery to Foata who happened to be a teaching visitor to the UCSD Mathematics Department that particular year. Foata did not seem particularly impressed by my imagery. To still obtain some satisfaction for my work, I told my idea to Viennot who happened to be visiting the UCSD Mathematics Department that same year. The rest is history. Viennot translated the rod idea into heaps of monomer–dimers to make a visual presentation of Flajolet combinatorialization of continued fractions and orthogonal polynomials. I would have expected Viennot to mention my visualization of the Cartier–Foata proof of the MacMahon Master theorem.

There is also a special feature that makes these Heaps notes particularly unique. Because my mathematical origins are mostly in analysis, I was already aware of the analytical theory of continued fraction and orthogonal polynomials. This given, before I presented Viennot's visualization of Cartier–Foata's work, I did at least state the basic theorems of the analytical theory. After that, I gave Viennot's purely combinatorial proofs of those same theorems so as to better appreciate the beauty and power of the combinatorial viewpoint.

Lecture 5: Basics on Finite Fields

When I started teaching elliptic curve cryptography in my cryptography course, I needed to have a reference to finite fields that I could recommend to my students and my teaching assistants. Unfortunately, all the books I examined were not elementary, voluminous, and contained also results that would be of interest only to specialists in their specific subjects. So I decided to write my own lecture notes. To help selecting what was important for my students to learn, I adopted the following criteria. I wanted a concise set of notes that contained only the basic theorems of finite fields. In addition, they needed to be elementary and easy to read.

My chosen sources were Berlekamp's *Algebraic Coding Theory* [4], Lidl and Niederreiter's *Introduction To Finite Fields and their Applications* [15], and Gaal's *Classical Galois Theory with Examples* [10].

Following the adopted criteria, the resulting notes now contain all the tools that are needed to computer explore finite fields, learn as much as possible about this branch of mathematics, and discover its magnificent beauty.

As I mentioned in the introduction, from the computational point of View, finite fields of the same dimension may actually have considerably different complexity. The reason is that every finite field is isomorphic to a quotient $\mathcal{F}_p[x]/(\phi(x))$, where $\phi(x)$ is a monic irreducible polynomial. One of the problems that is still unsolved is to determine for each finite field $\mathcal{F}$ of characteristic p, the simplest polynomial that generates $\mathcal{F}$ by the least complexity. Solutions of this problem should generate a variety of significant advances in the theory of finite fields.

San Diego, CA, USA   Adriano M. Garsia
Santa Barbara, CA, USA   Ömer Eğecioğlu
August 2020

Acknowledgements

It was noted by referees that Lecture 5 on Finite Fields is not aligned exactly with the subject matter of the others in this collection.

We would like to thank the editors who have decided to include it as a development in the spirit of exposition of the volume. We would also like to thank Ute McCrory and the editorial staff at Springer LNM for their assistance in the timely publication of this manuscript.

Acknowledgments

Contents

1 Alfred Young's Construction of the Irreducible Representations of S_n 1
   1.1   Introduction 1
   1.2   The Natural Representation Matrices 1
   1.3   Young's Tableau Idempotents 5
   1.4   Semisimplicity of Algebras 10
   1.5   Young's Matrix Units for $\mathcal{A}(\mathbf{S_n})$ 13
   1.6   Properties of the Representations $\{A^\lambda\}_\lambda$ and Their Characters 25
   1.7   The Frobenius Map 31

2 Young's Seminormal Representation, Murphy Elements, and Content Evaluations 35
   2.1   Introduction 35
   2.2   The Young Idempotents 36
   2.3   Young's Seminormal Units 42
   2.4   The Murphy Elements 55
   2.5   The Seminormal Matrices 63
   2.6   Murphy Elements and Conjugacy Classes 73

3 On Finite Dimensional $\mathfrak{sl}(2)$ Representations and an Application to Algebraic Combinatorics 97
   3.1   Basic Identities 97
   3.2   Diagonalizability of H 102
   3.3   Spernerity of $L[m, n]$ 123

4 Heaps, Continued Fractions, and Orthogonal Polynomials 137
   4.1   Introduction 137
   4.2   Heaps of Monomers and Dimers 137
   4.3   The Cartier–Foata Languages 140
   4.4   Orthogonal Polynomials and Continued Fractions 146
   4.5   Moments and Motzkin Paths 152
   4.6   Chebyshev Polynomials 164

4.7   The Rogers–Ramanujan Continued Fraction.......................... 168
4.8   Partitions and Hermite Polynomials 173
4.9   The Legendre Polynomials... 181
4.10   The Laguerre Polynomials 183

5   Finite Fields.. 187
5.1   Introduction 187
5.2   The Euclidean Algorithm 188
5.3   Polynomial Factorization.. 192
5.4   Cyclotomic Polynomials 197
5.5   The Frobenius Map 209
5.6   A Factorization Algorithm 221

Glossary 225

References... 229

Lecture 1
Alfred Young's Construction
of the Irreducible Representations of S_n

1.1 Introduction

Sections 1.2–1.6 of these lecture notes are dedicated to the work of Alfred Young
on the irreducible representation of the symmetric group S_n. Some simplifications
due to Von Neumann are also introduced. The effort that follows is dedicated to
Young's construction of the *matrix units*. In particular we obtain a proof that the
group algebra $\mathcal{A}(S_n)$ is semisimple. Young construction is so immediate to permit
the reader, after the first section, to write a program implementing the *Fourier
transform*. Using this tool the reader will be able to solve a variety of purely
algebraic problems about $\mathcal{A}(S_n)$. Section 1.7 introduces the *Frobenius map*, a most
valuable bridge translating representation theory problems into symmetric function
problems.

1.2 The Natural Representation Matrices

If λ is a partition of n (in symbols $\lambda \vdash n$), a filling of the (French) Ferrers diagram
of λ by the integers $1, 2, \ldots, n$ will be referred to as an *injective* tableau of shape λ.
The collection of all these tableaux will be denoted by $\mathrm{INJ}(\lambda)$. If $T \in \mathrm{INJ}(\lambda)$ and
the entries of T increase from left to right in the rows and from bottom to top on the
columns then T is said to be standard. The collection of all these tableaux will be
denoted by $ST(\lambda)$. We also write $\lambda(T) = \lambda$ for any $T \in \mathrm{INJ}(\lambda)$.

Given $T_1, T_2 \in ST(\lambda)$ we shall say that T_1 precedes T_2 in the Young *first letter
order* and write $T_1 <_{LL} T_2$ if the first entry of disagreement between T_1 and T_2 is
higher in T_1 than in T_2. For instance, in Fig. (1.2.1), T_1 and T_2 agree in the positions
of $1, 2, 3, 4, 5$. The first letter of disagreement is 6 and it is *higher* in T_1 than in T_2.
The positions of the remaining letters do not matter.

© The Editor(s) (if applicable) and The Author(s), under exclusive license
to Springer Nature Switzerland AG 2020
A. M. Garsia, Ö. Eğecioğlu, *Lectures in Algebraic Combinatorics*,
Lecture Notes in Mathematics 2277, https://doi.org/10.1007/978-3-030-58373-6_1

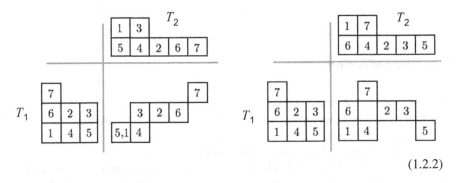

$$T_1 = \qquad\qquad T_2 = \qquad\qquad\qquad\qquad (1.2.1)$$

Given two tableaux T_1, T_2, not necessarily of the same shape, we let $T_1 \wedge T_2$ be the diagram obtained by placing in the cell (i, j) the intersection of row i of T_1 with column j of T_2. The diagram in Fig. (1.2.2) on the left gives an example of this construction

$$(1.2.2)$$

Note that in this case there is a cell which contains more than one entry. When this happens we say that $T_1 \wedge T_2$ is *bad*. Also note that in the example on the right in Fig. (1.2.2) each cell has at most one entry. When this happens we say that $T_1 \wedge T_2$ is *good*. In the good case, if $\lambda(T) = \lambda$ and $\lambda(T) = \mu$ then for each i we must have

$$\lambda_1 + \lambda_2 + \cdots + \lambda_i \ \leq \ \mu_1 + \mu_2 + \cdots + \mu_i . \qquad\qquad (1.2.3)$$

This is because the left hand side gives the number of entries in the first i rows of $T_1 \wedge T_2$ and the right hand side gives the number of entries in the tableau obtained by letting the entries of $T_1 \wedge T_2$ drop down their column until they are packed tight. This means that when T_1 and T_2 have different shapes, we must have some motion of cells, forcing a strict inequality in (1.2.3) for at least one i .

On the other hand when T_1 and T_2 have the same shape then $\lambda = \mu$, and there can be no motion of cells. In this case $T_1 \wedge T_2$ is also a tableau of shape $\lambda = \mu$, whose rows are a rearrangement of the rows of T_1 and whose columns are a rearrangement of the columns of T_2. To study with more detail the good case of two tableau T_1, $T_2 \in$ INJ(λ), we need further notation.

Given an injective tableau T with n cells, a permutation $\sigma = \sigma_1\sigma_2\ldots\sigma_n$ is made to act on T by replacing the entry i by σ_i. The resulting tableau is denoted by σT. We will use the symbol ϵ to denote the identity permutation. Of course then $\epsilon T = T$. We say that σ is in the *row group* of T if the rows of T and σT differ only by the order of their entries. The *column group* of T is analogously defined.

These two groups will be denoted by $R(T)$ and $C(T)$ respectively. This given, it is easy to see that when $\lambda = \mu$ and $T_1 \wedge T_2$ is good, then there are two permutations $\alpha_1 \in R(T_1)$ and $\beta_2 \in C(T_2)$ such that

$$T_1 \wedge T_2 = \alpha_1, \qquad\qquad T_2 = \beta_2\, T_1 \wedge T_2. \qquad\qquad (1.2.4)$$

Conversely, the existence of two such permutations giving (1.2.4) forces $T_1 \wedge T_2$ to be good.

This permits us to introduce an important function on pairs of tableau:

Definition 1.1 For $T_1, T_2 \in \mathrm{INJ}(\lambda)$ we let

$$C(T_1, T_2) = \begin{cases} 0 & \text{if } T_1 \wedge T_2 \text{ is bad,} \\ \mathrm{sign}(\beta_2) & \text{if } T_1 \wedge T_2 \text{ is good.} \end{cases}$$

Here and after, it will be convenient to let

$$S_1^\lambda, S_2^\lambda, \ldots, S_{f_\lambda}^\lambda \qquad\qquad (1.2.5)$$

denote the standard tableaux of shape λ in Young's first letter order. This given, we set

$$C^\lambda(\sigma) = \|C(S_i^\lambda, \sigma S_j^\lambda)\|_{i,j=1}^{f_\lambda}. \qquad\qquad (1.2.6)$$

Proposition 1.1 *The matrix $C^\lambda(\epsilon)$ is upper unitriangular and is therefore invertible over the integers.*

Proof Note that if $T_1 <_{LL} T_2$ and a is the first letter of disagreement, then there is a letter b such that a and b are in the same column of T_1 and in the same row of T_2. Note that $b = 4$ for the example given in our first display. In general b is the letter which lies at the intersection of the column of the shape λ in which a lies in T_1 with the row in which a lies in T_2. This means that $T_2 \wedge T_1$ is bad. So $T_1 <_{LL} T_2$ implies $C(T_2, T_1) = 0$. Thus from the definition of first letter order we derive that

$$C(S_i^\lambda, S_j^\lambda) = 0 \qquad\qquad \text{for all } i > j \ .$$

Since $C(S_i^\lambda, S_i^\lambda) = 1$ holds true trivially, our proof is complete. $\qquad\square$

This proposition allows us to define

$$A^\lambda(\sigma) = C^\lambda(\epsilon)^{-1}\, C^\lambda(\sigma). \qquad\qquad (1.2.7)$$

Our goal in these notes is to show that the collection of matrix functions

$$\{A^{\lambda(\sigma)}\}_{\lambda \vdash n} \qquad\qquad (1.2.8)$$

form a complete set of irreducible representations of S_n. It should be pointed out that for $f \in \mathcal{A}(S_n)$ we extend the definition in (1.2.7) by setting

$$A^\lambda(f) = \sum_{\sigma \in S_n} f(\sigma) A^\lambda(\sigma) .$$

Using our favorite symbolic manipulation software, such as MAPLE or MATH-EMATICA, we can implement the construction of the matrix $A^\lambda(f)$. This done, we can also implement the Fourier transform of S_n. That is the block diagonal matrix

$$\mathcal{F}_n(f) = \bigoplus_{\lambda \vdash n} A^\lambda(f). \tag{1.2.9}$$

We will show in the next few sections that the map $f \in \mathcal{A}(S_n) \to \mathcal{F}_n(f)$ is an algebra isomorphism. This implies that algebraic problems of $\mathcal{A}(S_n)$ are reduced to simple linear algebra problems.

Remark 1.1 The simplicity of the definition in (1.2.7) should be compared with the intricate constructions that characterize the treatments of the representations of the symmetric groups given in recent and past literature following the work of A. Young. Because of peculiarities of Young's style of writing many of his successors never bothered to read, let alone tried to understand, his beautiful constructions. For instance you will often see the name *Specht module* associated with representations which actually are those defined in (1.2.7). What is more amusing is that Young himself in QSA IV [28] with surprising premonition, described some alternate ways of constructing representations of S_n which include as particular cases all the constructions that followed, including those that are "*characteristic free*." It is fair to say that some of these later rediscoveries are based on a naive misunderstanding of Young's purely combinatorial arguments.

Problems

1. Show that for any injective tableau T we have $\alpha\beta = \epsilon$ with $\alpha \in R(T)$ and $\beta \in C(T)$ if and only if $\alpha = \beta = \epsilon$.
2. Let $T \in INJ(\lambda)$ with $\lambda \vdash n$. Show that $T \wedge \sigma T$ is good if and only if $\sigma = \beta_2\alpha_1$ with $\beta_2 \in C(\sigma T)$ and $\alpha_1 \in R(T)$. Show that this factorization is unique. That is show that if $\sigma = \beta_2'\alpha_1'$, with $\beta_2' \in C(\sigma T)$ and $\alpha_1' \in R(T)$ then then $\beta_1' = \beta_1$ and $\alpha_1' = \alpha_1$.
3. Construct all standard tableaux of shape $(3, 2, 1)$ in the Young first letter order. Then construct the matrix $A^{(3,2)}(\sigma)$ for

$$\sigma = \begin{bmatrix} 1\ 2\ 3\ 4\ 5 \\ 3\ 5\ 4\ 1\ 2 \end{bmatrix} .$$

4. Show that when λ is a *hook* that is $\lambda = (k, 1^{n-k})$ then the matrix $C^\lambda(\epsilon)$ is the identity.
5. The *left regular representation* of S_n is the matrix function defined by the action of S_n on itself. More precisely we set $\sigma \langle \tau_1, \tau_2, \ldots, \tau_{n!} \rangle = \langle \tau_1, \tau_2, \ldots, \tau_{n!} \rangle L(\sigma)$. Equivalently

$$L(\sigma) = \left\| L_{i,j}(\sigma) \right\|_{i,j=1}^{n!} = \left\| \chi(\sigma \tau_j = \tau_i) \right\|_{i,j=1}^{n!}. \qquad (1.2.10)$$

Use this representation to show that if the element $f \in \mathcal{A}(S_n)$ is nilpotent then $f|_\epsilon = 0$. Hint: Prove that the matrix $L(f)$ has zero trace.
6. Determine if the conjugacy class of permutations with cycle structure $(2, 1, 1)$ is an invertible element of $\mathcal{A}(S_4)$. You may use a computer implementation of the Fourier transform of S_4 to prove invertibility.

1.3 Young's Tableau Idempotents

For a given tableau T here and after we set

$$P(T) = \sum_{\alpha \in R(T)} \alpha, \qquad N(T) = \sum_{\beta \in C(T)} \text{sign}(\beta)\beta. \qquad (1.3.1)$$

If $T_1, T_2 \in \text{INJ}(\lambda)$ by $\sigma_{T_1 T_2}$ we denote the permutation that sends T_2 into T_1 that is

$$T_1 = \sigma_{T_1 T_2} T_2 .$$

This given, it is easy to see that we have

$$(a) \;\; P(\sigma T) = \sigma P(T) \sigma^{-1} \; ; \;\; N(\sigma T) = \sigma N(T)\sigma^{-1} \qquad \text{for all} \;\; \sigma \in S_n,$$
$$(b) \;\; \sigma_{T_1 T_2} P(T_2) = P(T_1)\sigma_{T_1 T_2}, \qquad\qquad\qquad (1.3.2)$$
$$(c) \;\; \sigma_{T_1 T_2} N(T_2) = N(T_1)\sigma_{T_1 T_2}.$$

We shall also use the shorthand notations

$$(a) \;\; E_{T_1 T_2} = N(T_1) \, \sigma_{T_1 T_2} \, P(T_2), \quad E_T = E_{TT} = N(T)P(T), \qquad (1.3.3)$$
$$(b) \;\; F_{T_1 T_2} = P(T_1) \, \sigma_{T_1 T_2} \, N(T_2), \quad F_T = F_{TT} = P(T)N(T).$$

Note that from (1.3.2) (b) and (c) we derive that

$$(a) \;\; E_{T_1 T_2} = E_{T_1}\sigma_{T_1 T_2} = \sigma_{T_1 T_2} E_{T_2}, \qquad (1.3.4)$$
$$(b) \;\; F_{T_1 T_2} = F_{T_1}\sigma_{T_1 T_2} = \sigma_{T_1 T_2} F_{T_2} .$$

Recalling that

$$S_1^\lambda, \; S_2^\lambda, \; \ldots, \; S_{f_\lambda}^\lambda \tag{1.3.5}$$

denote the standard tableaux of shape λ in Young's first letter order we set

$$E_i^\lambda = E_{S_i^\lambda} \; ; \quad E_{ij}^\lambda = E_{S_i^\lambda S_j^\lambda}$$

$$F_i^\lambda = F_{S_i^\lambda} \; ; \quad F_{ij}^\lambda = F_{S_i^\lambda S_j^\lambda} \tag{1.3.6}$$

$$\sigma_{ij}^\lambda = \sigma_{S_i^\lambda S_j^\lambda} \; .$$

When dealing with a fixed shape, to lighten our formulas, sometime we shall omit the superscript λ from our symbols. We should note that many identities and results involving the elements E_{ij}^λ hold also for the F_{ij}^λ's, with only minor modifications. In each case, to avoid unnecessary repetitions, we shall prove them for one set and leave it to the reader to derive the result for the other set.

The group algebra elements $P(T)$, $N(T)$ have truly remarkable properties. To prove them we need some auxiliary results.

Proposition 1.2 *For a pair of tableaux T_1 and T_2 of the same shape the following properties are equivalent*

(a) $T_1 \wedge T_2$ *is good*

(b) $\exists \alpha_1 \in R(T_1)$ *and* $\beta_2 \in C(T_2)$ *such that* $T_2 = \beta_2 \alpha_1 T_1$

(c) $\exists \alpha_1 \in R(T_1)$ *and* $\beta_1 \in C(T_1)$ *such that* $T_2 = \alpha_1 \beta_1 T_1$

(d) $\exists \alpha_2 \in R(T_2)$ *and* $\beta_2 \in C(T_2)$ *such that* $T_2 = \alpha_2 \beta_2 T_1$

Moreover we have

$$\mathrm{sign}(\beta_1) = \mathrm{sign}(\beta_2) \; . \tag{1.3.7}$$

Proof We have seen that (a) and (b) are equivalent. From (1.3.2) (a) we then get that (b) implies

$$N(T_1) = \alpha_1^{-1} \beta_2^{-1} N(T_2) \beta_2 \alpha_1 = \alpha_1^{-1} N(T_2) \alpha_1 \; .$$

Thus there is a $\beta_1 \in N(T_1)$ such that

$$\beta_1 = \alpha_1^{-1} \beta_2 \alpha_1 \; . \tag{1.3.8}$$

This gives that

$$T_2 = \beta_2 \alpha_1 \, T_1 = (\alpha_1 \beta_1 \alpha_1^{-1}) \alpha_1 \, T_1 = \alpha_1 \beta_1 \, T_1$$

which is (c). Note that (1.3.7) now follows from (1.3.8). Conversely suppose we have (c). Then we may write

$$N(T_2) = \alpha_1 \beta_1 N(T_1) \beta_1^{-1} \alpha_1^{-1} = \alpha_1 N(T_1) \alpha_1^{-1}$$

and we must necessarily have a $\beta_2 \in C(T_2)$ such that

$$\beta_2 = \alpha_1 \beta_1 \alpha_1^{-1} \, ,$$

giving

$$\sigma_{T_2 T_1} = \alpha_1 \beta_1 = \beta_2 \alpha_1 \, .$$

This shows the equivalence of (b) and (c). We are left to show that (d) is equivalent to (b). More precisely from (b) and (1.3.2) (a) we derive that

$$P(T_2) = \beta_2 \alpha_1 (T_1) \alpha_1^{-1} \beta_2^{-1} = \beta_2 P(T_1) \beta_2^{-1} \, .$$

Thus there must be an $\alpha_2 \in R(T_2)$ such that

$$\alpha_2 = \beta_2 \alpha_1 \beta_2^{-1}$$

and this gives

$$\sigma_{T_2 T_1} = \alpha_2 \beta_1 = \alpha_2 \beta_2 \, .$$

Thus (b) implies (d). We leave it to the reader to show that (d) implies (b) and complete the proof of the proposition.   □

Remark 1.2   It is important to note that the permutations $\alpha_1, \alpha_2, \beta_1, \beta_2$ of Proposition 1.2 are uniquely determined by T_1 and T_2. In fact for a given Tableau T we cannot have two pairs $\alpha', \alpha'' \in R(T)$ and $\beta', \beta'' \in C(T)$ such that

$$\alpha' \beta' = \alpha'' \beta'' \, .$$

Indeed this is equivalent to the existence of an $\alpha \in R(T)$ and a $\beta \in C(T)$ such that

$$\alpha = \beta$$

and this is easily seen to be impossible. Let us keep this fact in mind since we are going to make use of it several times in the future.

Before we proceed with the next result we need to make some conventions. Young used a very efficient notation to represent some elements of the group algebra of the symmetric group S_n. For a given subset

$$S = \{1 \le i_1 < i_2 < \cdots < i_k \le n\} \subseteq \{1, 2, \ldots, n\}$$

we let the symbol $[S]$ represent the sum of all elements of S_n that permute the elements of S and leave fixed all the elements of the complement of S in $\{1, 2, \ldots, n\}$. He also used $[S]'$ to denote the sum of the same permutations where each is multiplied by its sign. For instance, with this convention we see that for the tableau

$$T = \begin{array}{|c|c|c|c|}
\hline
2 & 3 & \multicolumn{1}{c}{} & \multicolumn{1}{c}{} \\
\cline{1-2}
5 & 7 & \multicolumn{1}{c}{} & \multicolumn{1}{c}{} \\
\cline{1-3}
6 & 9 & 10 & \multicolumn{1}{c}{} \\
\hline
11 & 4 & 8 & 1 \\
\hline
\end{array}$$

we may write

$$P(T) = [1, 4, 8, 11] \times [6, 9, 10] \times [5, 7] \times [2, 3] \,,$$

$$N(T) = [2, 5, 6, 11]' \times [3, 4, 7, 9]' \times [8, 10]' \,.$$

Remark 1.3 For a given $f \in \mathcal{A}(S_n)$, it will also be convenient to denote by f' the element of the group algebra obtained by replacing each permutation by the same permutation preceded by its sign. More precisely, for $f = \sum_{\sigma \in S_n} f(\sigma)\,\sigma$ we shall set

$$f' = \sum_{\sigma \in S_n} \text{sign}(\sigma) f(\sigma)\,\sigma$$

and refer to the operation $f \to f'$ as the *priming operator*. We should note that any identity involving elements of the group algebra of S_n generates a companion identity when we apply the priming operator to both sides. In this manner identities for the E_{ij}^λ's may be transformed into identities for the F_{ij}^λ's. In fact, it easy to see that

$$E_{T_1, T_2}{}' = \text{sign}(\sigma_{T_1, T_2})\, F_{T_1^\top, T_2^\top} \qquad (1.3.9)$$

where $T_1^\top$ and $T_2^\top$ denote the corresponding transposed tableaux, where transposing a tableau simply means reflecting it across the $45°$ diagonal emanating from its south-west corner. In particular we have

$$E_{ij}^{\lambda}{}' = \text{sign}(\sigma_{ij}^{\lambda}) \, F_{ij}^{\lambda'} \, , \tag{1.3.10}$$

where λ' denotes the partition conjugate to λ.

Remark 1.4 For two partitions

$$\lambda = (\lambda_1, \lambda_2, \ldots, \lambda_h) \, , \quad \mu = (\mu_1, \mu_2, \ldots, \mu_k)$$

we say that μ *dominates* λ and write

$$\lambda \leq_D \mu$$

if and only if

$$\lambda_1 + \lambda_2 + \cdots + \lambda_i \leq \mu_1 + \mu_2 + \cdots + \mu_i \qquad \text{for all } 1 \leq i \leq \min(h, k).$$

Thus our remarks at the beginning of Sect. 1.3 may be restated by writing

$$T_1 \wedge T_2 \quad good \quad \longrightarrow \quad \text{shape}(T_1) \leq_D \text{shape}(T_2). \tag{1.3.11}$$

Now it develops that the same implication holds true if we have

$$(a) \ \ P(T_1)N(T_2) \neq 0 \quad \text{or} \quad (b) \ \ N(T_2)P(T_1) \neq 0. \tag{1.3.12}$$

To see this note that, by definition, $T_1 \wedge T_2$ *bad* implies that there are two elements a, b that are in the same row of T_1 and the same column of T_2. This means that the transposition (a, b) is at the same time in $R(T_1)$ and $C(T_2)$. We thus have

$$P(T_1)N(T_2) = \big(P(T_1)(a, b)\big)N(T_2) = P(T_1)\big((a, b)N(T_2)\big) = -P(T_1)N(T_2) \, ,$$

and this contradicts (1.3.12) (a). Thus

$$P(T_1)N(T_2) \neq 0 \quad \longrightarrow \quad T_1 \wedge T_2 \ good \quad \longrightarrow \quad \text{shape}(T_1) \leq_D \text{shape}(T_2) \, . \tag{1.3.13}$$

We can proceed similarly and show that

$$N(T_2)P(T_1) \neq 0 \quad \longrightarrow \quad T_1 \wedge T_2 \ good \quad \longrightarrow \quad \text{shape}(T_1) \leq_D \text{shape}(T_2) \, . \tag{1.3.14}$$

1.4 Semisimplicity of Algebras

We have seen that the group algebra $\mathcal{A}(G)$ of a finite group G has two binary operations "$+$" (*addition*) and "$\times$" (*multiplication*). The addition is trivially identified. Multiplication of two group algebra elements f_1 and f_2 is defined by setting

$$f_1 \times f_2 = \sum_{\sigma \in G} \sigma \sum_{\tau_1 \in G} \sum_{\tau_2 \in G} f_1(\tau_1) f_2(\tau_2) \chi (\sigma = \tau_1 \tau_2)$$

These two operations are associative, more precisely they satisfy the left and right distributivity laws. That is for all $f, h, g \in \mathcal{A}(G)$ we have

$$(left)\ \ h \times (f+g) = h \times f + h \times g \qquad\qquad (right)\ \ (f+g) \times h = f \times h + g \times h$$

Moreover, since its elements are formal linear combinations of group elements, such as

$$f = \sum_{\gamma \in G} f(\gamma) \gamma$$

with f a complex valued function on G, we also have a *multiplication by a scalar* operation, defined by setting for any complex number c

$$cf = \sum_{\gamma \in G} cf(\gamma) \gamma \ .$$

This operation is clearly associative and distributive with respect to addition.

The group algebra has also an identity which we denote ϵ which is simply given by the identity element of the group G.

A structure with these properties with scalars in a field $\mathbf{K}$ is usually referred to as a $\mathbf{K}$-Algebra. The collection of $n \times n$ matrices with entries in $\mathbf{K}$, which we denote $M_n[\mathbf{K}]$, is the simplest example of a $\mathbf{K}$-Algebra. It easily seen $M_n[\mathbf{K}]$ is a n^2-dimensional vector space with a natural basis consisting of the $n \times n$ matrices $E_{i,j}$ with i, j entry equal to 1 and all other entries equal to zero. With this notation, for any matrix $A = \|a_{i,j}\|_{i,j=1}^{n}$ we can write

$$A = \sum_{i=1}^{n} \sum_{j=1}^{n} a_{i,j} E_{i,j} \ . \tag{1.4.1}$$

The $E_{i,j}$, which (using Young's terminology) we will refer to as *matrix units* are easily seen to satisfy the identities

$$E_{i,j} \times E_{r,s} = \begin{cases} 0_n & \text{if } j \neq r \\ E_{i,s} & \text{if } j = r \end{cases} \tag{1.4.2}$$

where 0_n here means the $n \times n$ matrix with all elements equal to 0.

The next simplest example is obtained by taking a direct sum of a finite number of simple matrix algebras. More precisely we will denote

$$\bigoplus_{l=1}^{k} \mathcal{M}_{n_l}[\mathbf{K}]$$

the vector space of $m \times m$ block matrices, (with $m = \sum_{s=1}^{k} n_s$) of block diagonal form

$$A = \begin{bmatrix} A^{(1)} & 0 & \cdots & 0 \\ 0 & A^{(2)} & \cdots & 0 \\ \vdots & \vdots & \ddots & \vdots \\ 0 & 0 & \cdots & A^{(k)} \end{bmatrix} = \bigoplus_{l=1}^{k} A^{(l)} \quad \left(\text{with } A^{(l)} \in \mathcal{M}_{n_l}[\mathbf{K}]\right) \tag{1.4.3}$$

It will be convenient to denote by $f_{i,j}^{(s)}$ the block matrix in (1.4.3) with

$$A^{(l)} = \begin{cases} 0_{n_l} & \text{if } l \neq s \\ E_{i,j}^{(n_s)} & \text{if } l = s \end{cases} \tag{1.4.4}$$

where here $E_{i,j}^{(n_s)}$ denotes the $n_s \times n_s$ matrix with i,j element 1 and all other elements 0. It is easily seen that the collection $\bigcup_{s=1}^{k} \{f_{i,j}^{(s)}\}_{i,j=1}^{n_s}$ is a basis of $\bigoplus_{i=1}^{k} \mathcal{M}_{n_i}[\mathbf{K}]$, since it is clearly an independent set and every element $A = \bigoplus_{l=1}^{k} A^{(l)}$ has an expansion of the form

$$A = \sum_{s=1}^{k} \sum_{i=1}^{n_s} \sum_{j=1}^{n_s} a_{i,j}^{(s)} f_{i,j}^{(s)}.$$

In particular we see that

$$\dim \bigoplus_{s=1}^{k} \mathcal{M}_{n_s}[\mathbf{K}] = \sum_{s=1}^{k} n_s^2.$$

Moreover, from (1.4.2) and (1.4.4) it follows that

$$f_{i,j}^l \times f_{r,s}^m = \begin{cases} 0_M & \text{if } l \neq m \\ 0_M & \text{if } l = m \text{ but } j \neq r \\ f_{i,s}^l & l = m \text{ and } j \neq r \end{cases} \qquad \left(M = \sum_{s=1}^{k} n_s \right). \qquad (1.4.5)$$

It is easily shown that only multiples of the identity matrix $I \in M[\mathbf{K}]$ commute with all the matrices in $M[\mathbf{K}]$. In particular, it follows that the center of $\bigoplus_{i=1}^{k} M_{n_i}[\mathbf{K}]$ (that is the subspace of the elements that commute with all the other elements) is spanned by the elements

$$f_s = \sum_{i=1}^{n_s} f_{i,i}^{(n_s)} \qquad (1 \leq s \leq k), \qquad (1.4.6)$$

which are none other than the block matrices in (1.4.3) with

$$A^{(l)} = \begin{cases} 0_{n_l} & \text{if } l \neq s \\ I^{(n_s)} & \text{if } i = s \end{cases}$$

where $I^{(n_s)}$ denotes the $n_s \times n_s$ identity matrix. This gives

$$\dim Center\left(\bigoplus_{i=1}^{k} M_{n_i}[\mathbf{K}] \right) = k. \qquad (1.4.7)$$

It is customary to call the algebras $\bigoplus_{i=1}^{k} M_{n_i}[\mathbf{K}]$ *semisimple*.

Frobenius' Fundamental Theorem of representation theory states that the group algebra of every finite group is isomorphic to a semisimple algebra.

More precisely Frobenius result assures that given a finite group G of order $N = |G|$ there are integers $\{n_s\}_{s=1}^{k}$ such that

$$N = \sum_{s=1}^{k} n_s^2 \qquad (1.4.8)$$

and a bijective map

$$\phi : \mathcal{A}(G) \longleftrightarrow \bigoplus_{i=1}^{k} M_{n_i}[\mathbf{C}] \qquad (1.4.9)$$

that respects addition, multiplication and multiplication by a complex number. In particular we deduce from (1.4.7) that the integer k must be equal to the dimension of the center $C(G)$ of the group group algebra of G. But we have seen that $C(G)$ is none other that the subspace of class functions. Thus in particular we derive that k in (1.4.8) and (1.4.9) must be none other than the number of conjugacy classes of G.

Of course in view of (1.4.5) the Frobenius result amounts to showing the existence in $\mathcal{A}(G)$ of a basis $\bigcup_{s=1}^{k} \{e_{i,j}^{(s)}\}_{i,j=1}^{n_s}$ satisfying the identities

$$
e_{i,j}^{l} \times e_{r,s}^{m} = \begin{cases} 0 & \text{if } l \neq m \\ 0 & \text{if } l = m \text{ but } j \neq r \\ e_{i,s}^{l} & \text{if } l = m \text{ and } j = r. \end{cases} \tag{1.4.10}
$$

This given, the desired map ϕ is simply obtained by setting

$$
\phi\, e_{i,j}^{l} = f_{i,j}^{l}
$$

and extending it by linearity to the rest of $\mathcal{A}(G)$.

Remarkably, Alfred Young, initially unaware of Frobenius work, set himself the task in a series of papers written in a span of 17 years, to produce a purely combinatorial construction of such matrix units for several reflection groups including all the symmetric groups. While doing so he completely upstaged some of Frobenius' work, who constructed the k elements f_s in (1.4.6) for S_n.

In the next few sections we will present Young's construction for the symmetric groups and derive some of its most significant properties.

1.5   Young's Matrix Units for $\mathcal{A}(\mathbf{S_n})$

This section is dedicated to the proof of the following remarkable result.

Theorem 1.1 (A.Young) *With $S_1^{\lambda}, S_2^{\lambda}, \ldots, S_{n_\lambda}^{\lambda}$ the standard tableaux of shape $\lambda \vdash n$ in Young's first letter order, we have a constant $h_\lambda \neq 0$ assuring that the group algebra element*

$$
\gamma_i^{\lambda} = \frac{1}{h_\lambda}\, N(S_i^{\lambda}) P(S_i^{\lambda}) \tag{1.5.1}
$$

is an idempotent. Moreover, setting $\sigma_{S_i^{\lambda}, S_j^{\lambda}} = \sigma_{ij}^{\lambda}$, the group algebra elements

$$
e_{ij}^{\lambda} = \gamma_i^{\lambda}\, \sigma_{ij}^{\lambda}\, (1 - \gamma_{j+1}^{\lambda})(1 - \gamma_{j+2}^{\lambda}) \cdots (1 - \gamma_{f_\lambda}^{\lambda}) \tag{1.5.2}
$$

do not vanish and satisfy the identities

$$e_{ij}^{\lambda} \, e_{rs}^{\mu} = \begin{cases} e_{is}^{\lambda} & \text{if } \lambda = \mu \text{ and } j = r \\ 0 & \text{otherwise .} \end{cases} \tag{1.5.3}$$

We will obtain our proof by combining a few, very simple. purely combinatorial properties of the group algebra elements γ_i^{λ} which we will state as separate propositions.

We will start with the following very beautiful fact [27] which considerably simplifies Young's original argument.

Proposition 1.3 (Von Neumann Sandwich Lemma) *For any element $f \in \mathcal{A}(S_n)$ and any $T \in \text{INJ}(\lambda)$ we have*

$$N(T) \, f \, P(T) = c_T(f) \, N(T) P(T), \tag{1.5.4}$$

with

$$c_T(f) = f \, P(T)N(T) \, |_{\epsilon} \, . \tag{1.5.5}$$

Proof Expanding f in the left hand side of (1.5.4) we get

$$N(T) \, f \, P(T) = \sum_{\sigma \in S_n} f(\sigma) \, N(T) \, \sigma \, P(T) \, . \tag{1.5.6}$$

Now note that using (1.3.2) (a) we can write

$$N(T) \, \sigma \, P(T) = N(T) \, P(\sigma T) \sigma \, .$$

That means that the only terms that survive in (1.5.6) are those for which $N(T) \, P(\sigma T) \neq 0$. However, as we have seen in Remark 1.4, this can happen only if $\sigma T \wedge T$ is good. But part (d) of Proposition 1.2 gives that we must have

$$T = \alpha \, \beta \, \sigma \, T$$

with $\alpha \in R(T)$ and $\beta \in C(T)$. In other words the only terms that survive in (1.5.6) are those coming from permutations σ of the form

$$\sigma = \beta^{-1}\alpha^{-1} \qquad \qquad (\text{with } \alpha \in R(T), \ \beta \in C(T))$$

Since any permutation that can be so expressed has a unique expression of this form we can rewrite (1.5.6) as

$$N(T) f P(T) = \sum_{\alpha \in R(T)} \sum_{\beta \in C(T)} f(\beta^{-1}\alpha^{-1}) N(T) \beta^{-1}\alpha^{-1} P(T) . \qquad (1.5.7)$$

But then the simple identity

$$N(T) \beta^{-1}\alpha^{-1} P(T) = \text{sign}(\beta) N(T) P(T)$$

reduces (1.5.7) to

$$N(T) f P(T) = \left(\sum_{\alpha \in R(T)} \sum_{\beta \in C(T)} \text{sign}(\beta) f(\beta^{-1}\alpha^{-1}) \right) N(T) P(T) ,$$

which is (1.5.4) with

$$c_T(f) = \sum_{\alpha \in R(T)} \sum_{\beta \in C(T)} \text{sign}(\beta) f(\beta^{-1}\alpha^{-1}) .$$

However the latter is but another way of writing (1.5.5). $\qquad \square$

Proposition 1.3 yields us the following fundamental fact.

Proposition 1.4 *For each $\lambda \vdash n$ we have a nonvanishing constant h_λ depending only on λ such that for all tableaux $T \in \text{INJ}(\lambda)$ we have, $E_T = N(T)P(T)$*

$$E_T^2 = h_\lambda E_T . \qquad (1.5.8)$$

Proof We simply use (1.5.4) with $f = P(T)N(T)$ and get

$$N(T) P(T)N(T) P(T) = h_\lambda N(T)P(T) ,$$

with

$$h_\lambda = P(T)N(T) P(T)N(T) |_\epsilon .$$

Next we need to show that h_λ is the same for all injective tableaux of shape λ. However this is a simple consequence of the fact that if $T_1 = \sigma T$ then $P(T_1) = \sigma P(T)\sigma^{-1}$ and $N(T_1) = \sigma N(T)\sigma^{-1}$, thus

$$P(T_1)N(T_1) P(T_1)N(T_1) |_\epsilon = \sigma P(T)N(T) P(T)N(T)\sigma^{-1} |_\epsilon$$
$$= P(T)N(T) P(T)N(T) |_\epsilon .$$

Finally we must show that $h_\lambda \neq 0$. But if $h_\lambda = 0$ the element E_T would be nilpotent. This in turn would imply that its image $L(E_T)$ by the left regular representation would be a nilpotent matrix. Since nilpotent matrices have trace zero and for any group algebra element $f = \sum_{\sigma \in S_n} f(\sigma)\sigma$ we have

$$\mathrm{trace} L(f) = \sum_{\sigma \in S_n} f(\sigma)\, \mathrm{trace} L(\sigma) = f(\epsilon)n!$$

the vanishing of h_λ would imply that $E_T \mid_\epsilon = 0$. Now this is absurd since, from Remark 1.2 we immediately derive that $E_T \mid_\epsilon = 1$. This completes our proof.   □

Remark 1.5 We should note that one of the simplifications to Young's arguments due to the Von Neumann Lemma is to avoid a direct identification of the constant h_λ as $n!/n_\lambda$. This is obtained in Young's work as the final result of gruesome brute force proof of the idempotency of the group algebra elements E_T/h_λ. In the present development the identification of h_λ will be carried out at the very end. This given, until that time we will assume that h_λ is the mysterious constant appearing in Von Neuman's lemma. In particular, this assumption yields us that all the elements γ_i^λ are idempotent.

Proposition 1.5

$$\gamma_j^\lambda \gamma_i^\mu = \begin{cases} 0 & \text{if } \lambda \neq \mu \\ 0 & \text{if } \lambda = \mu \text{ but } j > i \\ \gamma_i & \text{if } \lambda = \mu \text{ and } j = i \,. \end{cases} \qquad (1.5.9)$$

Proof We have seen in the proof of Proposition 1.1 that if $T_1 <_{LL} T_2$ then there is a pair of entries r, s that are in the same column of T_1 and same row of T_2 . This gives

$$P(T_2)N(T_1) = P(T_2)(r, s)N(T_1) = -P(T_2)N(T_1)$$

and thus $E_{T_2} E_{T_1}$ must necessarily vanish. Since $S_1^\lambda, S_2^\lambda, \ldots, S_{n_\lambda}^\lambda$ are in Young's first letter order it follows that

$$E_{S_j^\lambda} E_{S_i^\lambda} = 0 \quad \text{when } j > i$$

in view of (1.5.1) this proves the second case of (1.5.9).

To prove the first case of (1.5.9) we need only show that if T_1 and T_2 are injective tableaux of shapes λ and μ respectively then

$$\lambda \neq \mu \quad \longrightarrow \quad E_{T_1} E_{T_2} = 0 \,.$$

To this end note that (1.3.13) immediately gives that

$$E_{T_1} E_{T_2} = N(T_1)P(T_1)N(T_2)P(T_2) \neq 0 \quad \longrightarrow \quad \lambda \leq \mu . \tag{1.5.10}$$

On the other hand we expand the mid element $P(T_1)N(T_2)$ in the product $E_{T_1}E_{T_2}$ we will obtain a sum of terms of the form

$$N(T_1)\sigma P(T_2) = N(T_1)P(\sigma T_2)\sigma .$$

If $E_{T_1}E_{T_2}$ does not vanish then at least one of these terms must not vanish. But then from (1.3.14) we derive that $\mu \leq \lambda$ which together with (1.5.10) contradicts $\lambda \neq \mu$. This completes our proof of the first case of (1.5.9) and we are done since the last case, under our assumption that h_λ is the Von Neumannn constant, is as we have seen an immediate consequence of Proposition 1.4. $\qquad\square$

We are now finally in a position to prove Theorem 1.1, except for the identification of the constant h_λ. We will start with a proof of the identities in (1.5.3).

Proof of the Identities in (1.5.3)

Note first that if $\lambda \neq \mu$ then the first case of (1.5.9) gives

$$(1 - \gamma_i^\lambda)\gamma_j^\mu = \gamma_j^\mu \qquad \text{(for all } i, j\text{)} . \tag{1.5.11}$$

Thus

$$
\begin{aligned}
e_{ij}^\lambda e_{rs}^\mu &= \sigma_{ij}^\lambda \gamma_j^\lambda (1 - \gamma_{j+1}^\lambda) \cdots (1 - \gamma_{n_\lambda}^\lambda) \gamma_r^\mu \sigma_{rs}^\mu (1 - \gamma_{s+1}^\mu) \cdots (1 - \gamma_{n_\lambda}^\mu) \\
&= \sigma_{ij}^\lambda \gamma_j^\lambda \gamma_r^\mu \sigma_{rs}^\mu (1 - \gamma_{s+1}^\mu) \cdots (1 - \gamma_{n_\lambda}^\mu) \\
&= 0 \qquad \text{(again by the first case of (1.5.9))} .
\end{aligned}
$$

We are thus reduced to proving the identities in (1.5.3) when $\lambda = \mu$ in all our elements. Note then that the second case of (1.5.9) gives

$$\text{(a) } \gamma_j^\lambda \gamma_i^\lambda = 0 \quad \text{as well as} \quad \text{(b) } (1 - \gamma_j^\lambda)\gamma_i^\lambda = \gamma_i^\lambda \qquad \text{when } j > i . \tag{1.5.12}$$

Now note that we can write, for $r < j$

$$e_{ij}^\lambda e_{rs}^\lambda = \sigma_{ij}^\lambda \gamma_j^\lambda (1 - \gamma_{j+1}^\lambda) \cdots (1 - \gamma_{n_\lambda}^\lambda) \gamma_r^\lambda \sigma_{rs}^\lambda (1 - \gamma_{s+1}^\lambda) \cdots (1 - \gamma_{n_\lambda}^\lambda)$$

$$\text{(by (1.5.12) (b))} = \sigma_{ij}^\lambda \gamma_j^\lambda \gamma_r^\mu \sigma_{rs}^\mu (1 - \gamma_{s+1}^\mu) \cdots (1 - \gamma_{n_\lambda}^\mu)$$

$$\text{(by (1.5.12) (a))} = 0 ,$$

and for $r = j$

$$e_{ij}^\lambda e_{rs}^\lambda = \sigma_{ij}^\lambda \gamma_j^\lambda (1 - \gamma_{j+1}^\lambda)$$

$$\cdots (1 - \gamma_{n_\lambda}^\lambda) \gamma_j^\lambda \sigma_{js}^\lambda (1 - \gamma_{s+1}^\lambda) \cdots (1 - \gamma_{n_\lambda}^\lambda)$$

$$\text{(by (1.5.12) (a))} = \sigma_{ij}^\lambda \gamma_j^\lambda \gamma_j^\lambda \sigma_{js}^\lambda (1 - \gamma_{s+1}^\lambda) \cdots (1 - \gamma_{n_\lambda}^\lambda)$$

$$\text{(last case of (1.5.9))} = \sigma_{ij}^\lambda \gamma_j^\lambda \sigma_{js}^\lambda (1 - \gamma_{s+1}^\lambda) \cdots (1 - \gamma_{n_\lambda}^\lambda)$$

$$= \gamma_i^\lambda \sigma_{ij}^\lambda \sigma_{js}^\lambda (1 - \gamma_{s+1}^\lambda) \cdots (1 - \gamma_{n_\lambda}^\lambda) = e_{is}^\lambda .$$

Finally for $r > j$, after successive uses of (1.5.12) will necessarily get to a point where the resulting expression for $e_{ij}^\lambda e_{rs}^\lambda$ contains the factor $(1 - \gamma_r^\lambda)\gamma_r^\lambda$ which of course vanishes by the last case of (1.5.9). This completes our argument.

From these relations we can now establish the following fact.

Proposition 1.6 *The group algebra elements e_{ij}^λ cannot vanish.*

Proof Note first that the identity

$$e_{i,j}^\lambda e_{j,i}^\lambda = e_{i,i}^\lambda$$

reduces us to showing that the $e_{i,i}^\lambda$ themselves cannot vanish. Now this result is an immediate consequence of the very useful fact (as we shall see) that for any two group algebra elements f, g we have

$$fg\Big|_\epsilon = gf\Big|_\epsilon \tag{1.5.13}$$

the reader is urged to work out a proof of this identity in full generality. This given, note first that from the definition in (1.5.2) and (1.5.13) it follows that

$$e_{i,i}^\lambda \Big|_\epsilon = \gamma_i^\lambda (1 - \gamma_{i+1}^\lambda) \cdots (1 - \gamma_{n_\lambda}^\lambda)\Big|_\epsilon$$

$$= (1 - \gamma_{i+1}^\lambda) \cdots (1 - \gamma_{n_\lambda}^\lambda)\gamma_i^\lambda \Big|_\epsilon$$

$$\text{(by (1.5.12) (b))} = \gamma_i^\lambda \Big|_\epsilon = \frac{1}{h_\lambda} \neq 0 . \qquad \square$$

As a first step in proving that Young's matrix units are a basis we can now establish that

Proposition 1.7 *The group algebra elements $\bigcup_{\lambda \vdash n} \{e_{ij}^\lambda\}_{i,j=1}^{n_\lambda}$ are independent.*

Proof Assume if possible that for some constants c_{ij}^{λ} we have

$$\sum_{\lambda \vdash n} \sum_{i,j=1}^{n_{\lambda}} c_{ij}^{\lambda} \, e_{ij}^{\lambda} = 0 \, .$$

Then for all possible choices of λ, r and s we derive from the identities in (1.5.3) that

$$c_{rs}^{\mu} \, e_{rs}^{\mu} = e_{rr}^{\mu} \Big(\sum_{\lambda \vdash n} \sum_{i,j=1}^{n_{\nu}} c_{ij}^{\lambda} \, e_{ij}^{\lambda} \Big) e_{ss}^{\mu} = 0$$

and since, as we have seen, the e_{rs}^{μ}'s do not vanish we must necessarily have $c_{rs}^{\mu} = 0$ as desired.   □

To prove that an independent set in a vector space V is a basis we need only show that it spans V or that its cardinality equals the dimension of V. Alfred Young proved that his matrix units are a basis, both ways. We will follow the latter approach first since it based on two beautiful combinatorial identities.

But before we can state them we need a few preliminary observations. In Fig. (1.5.14) we have depicted the Ferrers diagram of the partition $(3, 1, 1)$ and its immediate neighbors in the Young lattice (the lattice of Ferrers diagrams ordered by inclusion).

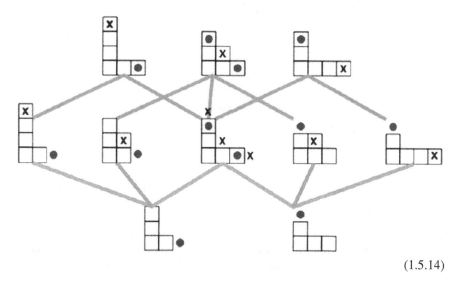

$$(1.5.14)$$

We have at the center the Ferrers diagram of $(3, 1, 1)$ with the circles indicating its *removable* (inner) corner cells and the $\mathbf{x}'s$ showing its *addable* (outer) corner cells. In the row below we have the partitions obtained by removing one of its inner corner cells and in the row above we have the partitions obtained by adding one of its addable outer corner cells. We can clearly see that every Ferrers diagram has one more addable cells than removable ones. In the same row as $(3, 1, 1)$ we have depicted the partitions that are obtained by going up a row by adding an addable cell then down a row by removing a removable cell. The fundamental property that yields our desired *basis* result is that the same collection of partitions can be obtained by first removing a removable cell then adding one of the addable cells. Using the symbol $\nu \to \mu$ to express that μ is obtained from ν by adding an addable cell of ν, or equivalently, ν is obtained from μ by removing one of the removable cells of μ, we can derive the following basic recursions for the number of standard tableaux.

Proposition 1.8 *Denoting by n_λ the number of standard tableaux of shape λ we have*

$$(a) \quad \sum_{\nu \vdash n-1} n_\nu \chi(\nu \to \mu) = n_\mu , \qquad (b) \quad \sum_{\lambda \vdash n+1} n_\lambda \chi(\mu \to \lambda) = (n+1)n_\mu .$$

$$(1.5.15)$$

Proof The identity in (1.5.15) (a) is immediate. In fact, if μ is a partition of n we obtain a bijection between the standard tableaux enumerated by the right hand side and the collection of standard tableau enumerated by the left hand side by the removal of n and the cell containing it. For (1.5.15) (b) we will proceed by induction on n. We can take as base case $\mu = (1)$ since we trivially see there is only one standard tableau of shape (1) and exactly one each of shapes $(1, 1)$ and (2). So let us assume that (1.5.15) (b) is true for $n-1$ and note that multiple applications of (1.5.15) (a) give

$$\sum_{\lambda \vdash n+1} n_\lambda \chi(\mu \to \lambda) = \sum_{\lambda \vdash n+1} \sum_{\gamma \vdash n} n_\gamma \chi(\mu \to \lambda \leftarrow \gamma) . \qquad (1.5.16)$$

Since in the sums on the right hand side we are allowed to remove the cell we are adding, we can see that, if μ has r inner corner cells (and therefore $r + 1$ outer corner sells), then by separating out the terms with $\gamma = \mu$ we can rewrite (1.5.16) in the form

$$\sum_{\lambda \vdash n+1} n_\lambda \chi(\mu \to \lambda) = \sum_{\lambda \vdash n+1} \sum_{\gamma \neq \mu} n_\gamma \chi(\mu \to \lambda \leftarrow \gamma) + (r+1)n_\mu . \qquad (1.5.17)$$

But the observation, that for any given μ, adding and addable and then removing a removable yields the same collection we obtain by the reverse process of removing a removable and then adding an addable, can be simply translated into the following beautiful identity

$$\sum_{\lambda \vdash n+1} \sum_{\gamma \neq \mu} n_\gamma \chi(\mu \to \lambda \leftarrow \gamma) = \sum_{\nu \vdash n-1} \sum_{\gamma \neq \mu} n_\gamma \chi(\mu \leftarrow \nu \to \gamma) . \qquad (1.5.18)$$

Thus, using (1.5.18) in (1.5.17) we derive that

$$\sum_{\lambda\vdash n+1} n_\lambda \chi(\mu\to\lambda) = \sum_{\nu\vdash n-1}\sum_{\gamma\neq\mu} n_\gamma \chi(\mu\leftarrow\nu\to\gamma) + (r+1)n_\mu . \qquad (1.5.19)$$

But now, since μ has r removable corners we see that the sum

$$\sum_{\gamma\vdash n} n_\gamma \chi(\mu\leftarrow\nu\to\gamma)$$

will contain exactly r terms where $\gamma = \mu$. This gives

$$\sum_{\nu\vdash n-1}\sum_{\gamma\vdash n} n_\gamma \chi(\mu\leftarrow\nu\to\gamma) = \sum_{\nu\vdash n-1}\sum_{\gamma\neq\mu} n_\gamma \chi(\mu\leftarrow\nu\to\gamma) + rn_\mu .$$

Using this (1.5.19) becomes

$$\sum_{\lambda\vdash n+1} n_\lambda \chi(\mu\to\lambda) = \sum_{\nu\vdash n-1}\sum_{\gamma\vdash n} n_\gamma \chi(\mu\leftarrow\nu\to\gamma) + n_\mu$$

$$= \sum_{\nu\vdash n-1} \chi(\nu\to\mu)\sum_{\gamma\vdash n} n_\gamma \chi(\nu\to\gamma) + n_\mu$$

$$\text{(by (1.5.15) (b) for } n-1) = \sum_{\nu\vdash n-1} \chi(\nu\to\mu)\, n\, n_\nu + n_\mu$$

$$\text{(by (1.5.15) (a))} = n\, n_\mu + n_\mu = (n+1)n_\mu ,$$

completing the induction and the proof.   □

These two recursions combined have the following remarkable consequence.

Proposition 1.9

$$\sum_{\lambda\vdash n} n_\lambda^2 = n! \qquad (1.5.20)$$

and in particular it follows that the group algebra elements $\bigcup_{\lambda\vdash n} \{e_{ij}^\lambda\}_{i,j=1}^{n_\lambda}$ *yield a basis for* $\mathcal{A}(S_n)$.

Proof Note that for the base case $n=2$ we have $n_{(1,1)}^2 + n_{(2)}^2 = 1+1 = 2!$. Thus we can again proceed by induction on n. We will suppose then that (1.5.20) is true for n. Now, using (1.5.15) (a), the left hand side of (1.5.20) for $n + 1$ becomes

$$\sum_{\lambda \vdash n+1} n_\lambda^2 = \sum_{\lambda \vdash n+1} n_\lambda \sum_{\mu \vdash n} n_\mu \chi(\mu \to \lambda)$$

$$= \sum_{\mu \vdash n} n_\mu \sum_{\lambda \vdash n+1} n_\lambda \chi(\mu \to \lambda)$$

$$\left(\text{by } (1.5.15)\,(b)\right) = \sum_{\mu \vdash n} n_\mu \,(n + 1)\, n_\mu$$

$$\left(\text{by induction}\right) = (n + 1)\, n! = (n + 1)!\,,$$

completing the induction and the proof. □

It is good to keep in mind that Young's matrix units satisfy the following useful identities

Proposition 1.10 *For all λ we have*

$$\left. e_{i,j}^\lambda \right|_\epsilon = \begin{cases} 0 & \text{if } i \neq j \\ \left. \gamma_i^\lambda \right|_\epsilon = \frac{1}{h_\lambda} & \text{if } i = j\,. \end{cases} \tag{1.5.21}$$

Proof These identities are immediate consequences of (1.5.13). In fact, to begin with we obtain

$$\left. e_{i,j}^\mu \right|_\epsilon = \left. e_{i,1}^\mu e_{1,j}^\mu \right|_\epsilon = \left. e_{1,j}^\mu e_{i,1}^\mu \right|_\epsilon = \begin{cases} 0 & \text{if } i \neq j \\ \left. e_{11}^\lambda \right|_\epsilon & \text{if } i = j\,. \end{cases}$$

But on the other hand we have

$$\left. e_{11}^\lambda \right|_\epsilon = \left. \gamma_1^\lambda (1 - \gamma_2^\lambda) \cdots (1 - \gamma_{n_\lambda}^\lambda) \right|_\epsilon = \left. (1 - \gamma_2^\lambda) \cdots (1 - \gamma_{n_\lambda}^\lambda) \gamma_1^\lambda \right|_\epsilon$$

$$= \left. \gamma_1^\lambda \right|_\epsilon = \frac{1}{h_\lambda} \left. E_{S_1^\lambda} \right|_\epsilon = \frac{1}{h_\lambda}$$

as desired. □

This given, it develops that there is a very simple explicit formula yielding the expansion of any element of the group algebra of S_n in terms of Young's matrix units.

Theorem 1.2 *For any $f \in \mathcal{A}(S_n)$ we have*

$$f = \sum_{\lambda \vdash n} h_\lambda \sum_{i,j=1}^{n_\lambda} \left. f \times e_{j,i}^\lambda \right|_\epsilon e_{i,j}^\lambda\,. \tag{1.5.22}$$

Proof Since the collection $\bigcup_{\lambda \vdash n} \{e_{ij}^{\lambda}\}_{i,j=1}^{n_{\lambda}}$ is a basis there will be coefficients $c_{ij}^{\lambda}(f)$ yielding

$$f = \sum_{\lambda \vdash n} \sum_{i,j=1}^{n_{\lambda}} c_{i,j}^{\lambda}(f) \, e_{i,j}^{\lambda} \, .$$

Thus

$$f \times e_{rs}^{\mu} \Big|_{\epsilon} = \sum_{i,j=1}^{n_{\mu}} c_{i,j}^{\mu}(f) \, e_{i,j}^{\mu} e_{rs}^{\mu} \Big|_{\epsilon} = \sum_{i=1}^{n_{\mu}} c_{i,r}^{\mu}(f) \, e_{i,s}^{\mu} \Big|_{\epsilon}$$

and (1.5.20) gives

$$f \times e_{rs}^{\mu} \Big|_{\epsilon} = c_{s,r}^{\mu}(f) \frac{1}{h_{\lambda}} \, .$$

This proves (1.5.22). $\square$

Note that, if we interpret every permutation $\sigma \in S_n$ as the element of $\mathcal{A}(S_n)$ which is 1 on σ and 0 elsewhere, we will have coefficients $a_{i,j}^{\lambda}(\sigma)$ yielding the expansion

$$\sigma = \sum_{\lambda \vdash n} \sum_{i,j=1}^{n_{\lambda}} a_{i,j}^{\lambda}(\sigma) \, e_{i,j}^{\lambda} \, . \tag{1.5.23}$$

Now we have the following remarkable fact.

Theorem 1.3 *For each $\lambda \vdash n$ the matrices $\{A^{\lambda}(\sigma) = \|a_{i,j}^{\lambda}(\sigma)\|_{i,j=1}^{n_{\lambda}}\}_{\sigma \in S_n}$ yield an irreducible representation of S_n.*

Proof For any $\alpha, \beta \in S_n$ we have, using (1.5.23) and the identities in (1.5.3)

$$\alpha\beta = \sum_{\lambda \vdash n} \sum_{i,j=1}^{n_{\lambda}} \sum_{r,s=1}^{n_{\lambda}} a_{i,j}^{\lambda}(\alpha) a_{r,s}^{\lambda}(\beta) \, e_{i,j}^{\lambda} e_{r,s}^{\lambda}$$

$$= \sum_{\lambda \vdash n} \sum_{i,j=1}^{n_{\lambda}} \sum_{s=1}^{n_{\lambda}} a_{i,j}^{\lambda}(\alpha) a_{j,r}^{\lambda}(\beta) \, e_{i,s}^{\lambda} \tag{1.5.24}$$

$$= \sum_{\lambda \vdash n} \sum_{i=1}^{n_{\lambda}} \sum_{s=1}^{n_{\lambda}} \left(\sum_{j=1}^{n_{\lambda}} a_{i,j}^{\lambda}(\alpha) a_{j,r}^{\lambda}(\beta) \right) e_{i,s}^{\lambda} \, .$$

But on the other hand a direct use of (1.5.23) for $\sigma = \alpha\beta$ gives

$$\alpha\beta = \sum_{\lambda \vdash n} \sum_{i,j=1}^{n_\lambda} a_{i,j}^\lambda(\alpha\beta)\, e_{i,j}^\lambda .$$

Comparing with (1.5.24) yields the desired product identity

$$A^\lambda(\alpha\beta) = A^\lambda(\alpha)A^\lambda(\beta) . \tag{1.5.25}$$

Note further that an application of (1.5.22) for $f = \epsilon$ yields, (by (1.5.20))

$$\epsilon = \sum_{\lambda \vdash n} h_\lambda \sum_{i,j=1}^{n_\lambda} e_{j,i}^\lambda \Big|_\epsilon e_{i,j}^\lambda = \sum_{\lambda \vdash n} \sum_{i=1}^{n_\lambda} e_{i,i}^\lambda .$$

In other words we also have $A^\lambda(\epsilon) = I_{n_\lambda}$, the $n_\lambda \times n_\lambda$ identity matrix!.

Next note that using (1.5.23) we derive that

$$\sigma e_{r,s}^\mu = \sum_{i,j=1}^{n_\mu} a_{i,j}^\mu(\sigma)\, e_{i,j}^\mu e_{r,s}^\mu = \sum_{i=1}^{n_\mu} a_{i,r}^\mu(\sigma)\, e_{i,s}^\mu . \tag{1.5.26}$$

In matrix form, this can be written as

$$\sigma\big(e_{1,s}^\mu, e_{2,s}^\mu, \ldots, e_{n_\mu,s}^\mu\big) = \big(e_{1,s}^\mu, e_{2,s}^\mu, \ldots, e_{n_\mu,s}^\mu\big)A^\mu(\sigma) . \tag{1.5.27}$$

In other words the have shown that the subspace $\mathcal{L}[e_{1,s}^\mu, e_{2,s}^\mu, \ldots, e_{n_\mu,s}^\mu]$ spanned by the independent set $\{e_{1,s}^\mu, e_{2,s}^\mu, \ldots, e_{n_\mu,s}^\mu\}$ is invariant under the action of S_n and the matrix corresponding to the action of σ on this basis is precisely $A^\mu(\sigma)$.

To show that $A^\mu(\sigma)$ is an irreducible representation we need to show that $\mathcal{L}[e_{1,s}^\mu, e_{2,s}^\mu, \ldots, e_{n_\mu,s}^\mu]$ has no invariant subspace other than $\{0\}$. In fact, we can show that the S_n action on any nonvanishing element $g \in \mathcal{L}[e_{1,s}^\mu, e_{2,s}^\mu, \ldots, e_{n_\mu,s}^\mu]$, can yield any one of the basis elements $\{e_{1,s}^\mu, e_{2,s}^\mu, \ldots, e_{n_\mu,s}^\mu\}$.

To this end suppose that

$$g = \sum_{i=1}^{n_\mu} c_i e_{i,s}^\mu .$$

Since $g \neq 0$ there must be at least one coefficient, say $c_u \neq 0$. This given we have

$$\frac{1}{c_u} e_{r,u}^\mu g = \sum_{i=1}^{n_\mu} \frac{c_i}{c_u} e_{ru}^\mu e_{i,s}^\mu = e_{r,s}^\mu$$

and the arbitrariness of r proves the irreducibility, completing our proof. $\square$

1.6 Properties of the Representations $\{A^\lambda\}_\lambda$ and Their Characters

To identify Young's matrices $A^\lambda(\sigma)$ and their traces we will need an auxiliary fact which reveals a beautiful property of the tableaux function $C(T_1, T_2)$.

Proposition 1.11 *For any $T_1, T_2 \in \text{INJ}(\lambda)$ and $\tau \in S_n$*

$$E_{T_2 T_1}\Big|_\tau = C(\tau T_1, T_2) . \tag{1.6.1}$$

Proof

$$E_{T_2 T_1}\Big|_\tau = \sum_{\beta_2 \in C(T_2)} \sum_{\alpha_2 \in R(T_2)} \text{sign}(\beta_2)\, \beta_2 \alpha_2\, \sigma_{T_2 T_1}\Big|_\tau$$

$$= \sum_{\beta_2 \in C(T_2)} \sum_{\alpha_2 \in R(T_2)} \text{sign}(\beta_2)\, \chi\big(\beta_2 \alpha_2\, \sigma_{T_2, T_1} = \tau\big) .$$

Now there may not be a single pair α_2, β_2 for which $\beta_2 \alpha_2\, \sigma_{T_2, T_1} = \tau$, in which case

$$E_{T_2 T_1}\,|_\tau = 0 .$$

But if there is such a pair, there will be one and only one as we noted in Remark 1.2. Then we must have

$$\beta_2 \alpha_2\, \sigma_{T_2 T_1} = \tau \quad \rightarrow \quad \beta_2 \alpha_2 = \tau\, \sigma_{T_1 T_2} = \sigma_{\tau T_1, T_2}$$

or

$$\sigma_{T_2, \tau T_1} = \alpha_2^{-1} \beta_2^{-1}$$

which combined with part (d) of Proposition 1.2 gives that $\tau T_1 \wedge T_2$ is good. Thus from (1.3.7) we derive that

$$E_{T_2 T_1}\,|_\tau = \begin{cases} 0 & \text{if } \tau T_1 \wedge T_2 \text{ is bad} \\ \text{sign}(\beta_2) & \text{if } \tau T_1 \wedge T_2 \text{ is good} \end{cases}$$

or

$$E_{T_2 T_1}\,|_\tau = C(\tau T_1, T_2)$$

as desired. $\square$

We should note that in particular we must have

$$E_{T_2 T_1} |_\epsilon = \text{sign}(\beta_2) = C(T_1, T_2).$$
(1.6.2)

This brings us in a position to prove the following theorem.

Theorem 1.4 *For any injective tableau T of shape $\mu \vdash n$ we have coefficients $a_i(T)$ giving*

$$E_T = \sum_{i=1}^{n_\mu} a_i(T) E_{S_i^\mu, T}.$$
(1.6.3)

In fact these coefficients may be directly obtained from the vector identity

$$\begin{pmatrix} a_1(T) \\ a_2(T) \\ \vdots \\ a_{n_\mu}(T) \end{pmatrix} = C^\mu(\epsilon)^{-1} \begin{pmatrix} C(S_1^\mu, T) \\ C(S_2^\mu, T) \\ \vdots \\ C(S_{n_\mu}^\mu, T) \end{pmatrix}.$$
(1.6.4)

Proof Assume first that these coefficients do exist. Multiplying (1.6.3) on the right by $\sigma_{T S_j^\mu}$ gives

$$E_{T S_j^\mu} = \sum_{i=1}^{n_\mu} a_i(T) E_{S_i^\mu S_j^\mu}.$$

Equating coefficients of the identity, and using (1.6.2), we get

$$C(S_j^\mu, T) = \sum_{i=1}^{n_\mu} a_i(T) C(S_j^\mu, S_i^\mu)$$

which, in matrix notation, may be expressed as

$$\begin{pmatrix} c(S_1^\mu, T) \\ c(S_2^\mu, T) \\ \vdots \\ c(S_{n_\mu}^\mu, T) \end{pmatrix} = C^\mu(\epsilon) \begin{pmatrix} a_1(T) \\ a_2(T) \\ \vdots \\ a_{n_\mu}(T) \end{pmatrix},$$

and formula (1.6.4) follows upon left multiplication of both sides of this identity by $C^\lambda(\epsilon)^{-1}$.

It turns out that existence is an immediate consequence of (1.5.23), namely the general expansion identity

$$\sigma = \sum_{\lambda \vdash n} \sum_{i,j=1}^{n_\lambda} a_{i,j}^\lambda(\sigma) \, e_{i,j}^\lambda . \tag{1.6.5}$$

In fact, for any given μ and $1 \le r \le n_\mu$ the identities in (1.5.3) give (using (1.6.5))

$$\sigma \, \gamma_r^\mu = \sum_{i,j=1}^{n_\mu} a_{i,j}^\mu(\sigma) \, e_{i,j}^\mu \gamma_r^\mu = \sum_{i,j=1}^{n_\mu} a_{i,j}^\mu(\sigma) \, \gamma_i^\mu \sigma_{i,j}^\mu (1 - \gamma_j^\mu) \cdots (1 - \gamma_{n_\mu}^\mu) \gamma_r^\mu$$

$$= \sum_{i=1}^{n_\mu} a_{i,r}^\mu(\sigma) \, \gamma_i^\mu \sigma_{i,r}^\mu \gamma_r^\mu = \sum_{i=1}^{n_\mu} a_{i,r}^\mu(\sigma) \, \gamma_i^\mu \sigma_{i,r}^\mu .$$

Recalling the definition in (1.5.1), we derive from this that

$$\sigma \, E_{S_r^\mu} = \sum_{i=1}^{n_\mu} a_{i,r}^\mu(\sigma) \, E_{S_i^\mu, S_r^\mu} ,$$

which can be rewritten as

$$E_{\sigma S_r^\mu} = \sum_{i=1}^{n_\mu} a_{i,r}^\mu(\sigma) \, E_{S_i^\mu, S_r^\mu} \, \sigma^{-1}$$

and for $\sigma = \sigma_{T, S_r^\mu}$ we get

$$E_T = \sum_{i=1}^{n_\mu} a_{i,r}^\mu(\sigma_{T, S_r^\mu}) \, E_{S_i^\mu, S_r^\mu} \, \sigma_{S_r^\mu, T} = \sum_{i=1}^{n_\mu} a_{i,r}^\mu(\sigma_{T, S_r^\mu}) \, E_{S_i^\mu, T} .$$

This proves (1.6.3) with

$$a_i(T) = a_{i,r}^\mu(\sigma_{T, S_r^\mu}) \tag{1.6.6}$$

and completes our argument. $\qquad \square$

This proof has a most desirable by-product.

Theorem 1.5 *For all $\mu \vdash n$ and all $\sigma \in S_n$ we have*

$$A^\mu(\sigma) = C^\mu(\epsilon)^{-1} C^\mu(\sigma) . \tag{1.6.7}$$

Proof From (1.6.6) and (1.6.4) it follows for $\sigma S_r^\mu = T$ that

$$
\begin{pmatrix} a_{1,r}^\mu(\sigma) \\ a_{2,r}^\mu(\sigma) \\ \vdots \\ a_{n_\mu,r}^\mu(\sigma) \end{pmatrix} = C^\mu(\epsilon)^{-1} \begin{pmatrix} (S_1^\mu, \sigma S_r^\mu) \\ c(S_2^\mu, \sigma S_r^\mu) \\ \vdots \\ c(S_{n_\mu}^\mu, \sigma S_r^\mu) \end{pmatrix}
$$

But this simply says that the rth column of the matrix on the left of (1.6.7) equals the rth of the matrix on the right. Finally yielding the result which we have anticipated since the beginning of these notes.   □

We are now in a position to obtain Young's formula giving the characters of his irreducible representations. This result may be stated as follows.

Theorem 1.6 *For a given $\mu \vdash n$ set*

$$
\chi^\mu(\sigma) = \text{trace } A^\mu(\sigma) \quad \text{and} \quad \chi^\mu = \sum_{\sigma \in S_n} \chi^\mu(\sigma)\sigma . \tag{1.6.8}
$$

Then

$$
\chi^\mu = \sum_{T \in \text{INJ}(\mu)} \frac{P(T)(N(T)}{n!/n_\mu} . \tag{1.6.9}
$$

Proof Using (1.6.7) and dropping some of the μ superscripts we can write

$$
\chi^\mu = \sum_{\sigma \in S_n} \sigma \text{ trace } C(\epsilon)^{-1} C(\sigma) = \text{trace } C(\epsilon)^{-1} F \tag{1.6.10}
$$

where $F = \|F_{ij}\|_{i=1}^{n_\mu}$ is a matrix with group algebra entries

$$
F_{ij} = \sum_{\sigma \in S_n} \sigma \, C(S_i, \sigma S_j) .
$$

Now Proposition 1.11 gives that

$$
C(S_i, \sigma S_j) = C(\sigma^{-1} S_i, S_j) = E_{S_j S_i} |_{\sigma^{-1}} .
$$

Thus

$$
F_{ij} = \sum_{\sigma \in S_n} E_{S_j S_i} |_{\sigma^{-1}} \sigma = \sum_{\sigma \in S_n} E_{S_j S_i} |_\sigma \sigma^{-1} = P(S_i)\sigma_{ij} N(S_j) .
$$

Note further that since χ^μ is a class function it follows that we also have

$$\chi^\mu = \frac{1}{n!} \sum_{\sigma \in S_n} \sigma \chi^\mu \sigma^{-1}$$

and we can thus write, using (1.6.10)

$$\chi^\mu = \frac{1}{n!} \sum_{\sigma \in S_n} \sigma \left(\text{trace } C(\epsilon)^{-1} F \right) \sigma^{-1} = \frac{1}{n!} \text{ trace } C(\epsilon)^{-1} \sum_{\sigma \in S_n} \sigma F \sigma^{-1} .$$

But we see that

$$\sum_{\sigma \in S_n} \sigma F_{ij} \sigma^{-1} = \sum_{\sigma \in S_n} \sigma P(S_i) \sigma_{ij} N(S_j) \sigma^{-1}$$

$$= \sum_{\sigma \in S_n} \sigma N(S_j) P(S_i) \sigma_{ij} \sigma^{-1} = 0$$

at least when $i > j$. This means that the matrix

$$\sum_{\sigma \in S_n} \sigma F \sigma^{-1}$$

is upper triangular and since $C(\epsilon)^{-1}$ is upper triangular with unit diagonal elements we must conclude that

$$\chi^\mu = \frac{1}{n!} \text{ trace } \sum_{\sigma \in S_n} \sigma F \sigma^{-1} = \frac{1}{n!} \sum_{\sigma \in S_n} \sigma \left(\sum_{i=1}^{n_\mu} F_{i,i} \right) \sigma^{-1}$$

$$= \frac{1}{n!} \sum_{i=1}^{n_\mu} \sum_{\sigma \in S_n} \sigma P(S_i) N(S_i) \sigma^{-1} = \frac{n_\mu}{n!} \sum_{T \in \text{INJ}(\lambda)} P(T) N(T)$$

which is (1.6.9) precisely as asserted. □

The character χ^μ is in fact very closely related to the matrix units. More precisely

Proposition 1.12 *Setting*

$$e^\mu = \sum_{i=1}^{n_\mu} e_{i,i}^\mu \tag{1.6.11}$$

we have

$$\chi^\mu = h_\mu e^\mu \tag{1.6.12}$$

and in particular it follows from the identities in (1.5.3) that

$$\chi_\mu \times \chi_\mu = h_\mu \, \chi_\mu \, . \tag{1.6.13}$$

Proof Notice that using (1.5.23) we get

$$e^\mu_{i,i}\Big|_{\sigma^{-1}} = \sigma \, e^\mu_{i,i}\Big|_\epsilon = \sum_{\lambda \vdash n} \sum_{r,s=1}^{n_\lambda} a^\lambda_{r,s}(\sigma) \, e^\lambda_{r,s} \, e^\mu_{i,i}\Big|_\epsilon = \sum_{r,s=1}^{n_\mu} a^\mu_{r,s}(\sigma) \, e^\mu_{r,s} \, e^\mu_{i,i}\Big|_\epsilon$$

$$= \sum_{r=1}^{n_\mu} a^\mu_{r,i}(\sigma) \, e^\mu_{r,i}\Big|_\epsilon = a^\mu_{i,i}(\sigma)\tfrac{1}{h_\mu} \quad \text{(by (1.5.20))}$$

and summing for i

$$e^\mu\Big|_{\sigma^{-1}} = \sum_{i=1}^{n_\mu} e^\mu_{i,i}\Big|_{\sigma^{-1}} = \tfrac{1}{h_\mu}\chi^\mu(\sigma) \, . \tag{1.6.14}$$

Now note that since e^μ is a central element of $\mathcal{A}(S_n)$ it follows that

$$e^\mu\Big|_{\sigma^{-1}} = e^\mu\Big|_\sigma \, .$$

Thus multiplying both sides of (1.6.14) by σ and summing over S_n gives (1.6.12) and completes our proof.   □

Now a result parallel to (1.6.13) can be obtained by direct manipulations of Young idempotents.

Proposition 1.13

$$\left(\sum_{T \in \text{INJ}(\mu)} N(T)P(T) \right)^2 = h_\mu^2 \sum_{T \in \text{INJ}(\mu)} N(T)P(T) \, . \tag{1.6.15}$$

Proof Note we may rewrite the left hand aside of (1.6.15) as

$$LHS = \sum_{T \in \text{INJ}(\mu)} \sum_{\sigma \in S_n} N(T)P(T)N(\sigma T)P(\sigma T) \, . \tag{1.6.16}$$

But $P(T)N(\sigma T) \neq 0$ implies $T \wedge \sigma T$ is good. Thus from Proposition 1.2 we derive that the only nonvanishing summands in are obtained for $\sigma = \alpha\beta$ with $\alpha \in R(T)$ and $\beta \in C(T)$. Since we have

$$N(T)P(T)N(\alpha\beta T)P(\alpha\beta T) = N(T)P(T)\alpha\beta N(T)P(T)\beta^{-1}\alpha^{-1}$$

$$= N(T)P(T)N(T)P(T)\,\text{sign}(\beta)\beta^{-1}\alpha^{-1}$$

$$\text{(by (1.5.8))} = h_\mu N(T)P(T)\,\text{sign}(\beta)\beta^{-1}\alpha^{-1}$$

the identity in (1.6.16) becomes

$$
LHS = h_\mu \sum_{T \in \text{INJ}(\mu)} N(T)P(T) \sum_{\substack{\alpha \in R(T) \\ \beta \in C(T)}} \text{sign}(\beta)\beta^{-1}\alpha^{-1}
$$

$$
= h_\mu \sum_{T \in \text{INJ}(\mu)} N(T)P(T)N(T)P(T)
$$

and a further use of (1.5.8) gives (1.6.15) as desired.   □

The result we have anticipated since Von Neumann Lemma is now finally within reach.

Theorem 1.7

$$
h_\mu = \frac{n!}{n_\mu} . \tag{1.6.17}
$$

Proof Squaring both sides of (1.6.9) gives, using (1.6.13) and (1.6.15)

$$
h_\mu \chi^\mu = \frac{h_\mu^2}{(n!/n_\mu)^2} \sum_{T \in \text{INJ}(\mu)} P(T)N(T) .
$$

Canceling the common factor h_μ and using (1.6.9) again gives

$$
\chi^\mu = \frac{h_\mu}{n!/n_\mu} \chi^\mu
$$

and (1.6.17) follows from the nonvanishing of χ^μ.   □

1.7   The Frobenius Map

The group algebra $\mathcal{A}(S_n)$ has a natural scalar product obtained by setting for any $f, g \in \mathcal{A}(S_n)$

$$
< f, g > = \frac{1}{n!} \sum_{\gamma \in S_n} f(\gamma)\overline{g(\gamma)} . \tag{1.7.1}
$$

Note that since for all $\alpha \in S_n$

$$
\alpha f = \sum_{\gamma \in S_n} f(\gamma)\alpha\gamma = \sum_{\beta \in S_n} f(\alpha^{-1}\beta)\beta ,
$$

we have

$$< \alpha f, g > = \frac{1}{n!} \sum_{\beta \in S_n} f(\alpha^{-1}\beta)\overline{g(\beta)}$$

$$= \frac{1}{n!} \sum_{\gamma \in S_n} f(\gamma)\overline{g(\alpha\sigma)} = < f, \alpha^{-1}g > . \qquad (1.7.2)$$

Similarly we show that

$$< f\alpha, g > = < f, g\alpha^{-1} > . \qquad (1.7.3)$$

Let us denote by C_μ the element of the group algebra of S_n that is obtained by summing all the permutations with cycle structure μ. Now we have shown that the cardinality of this collection of permutations is given by the ratio

$$|C_\mu| = \frac{n!}{z_\mu} \qquad (1.7.4)$$

where for a μ with α parts equal to i it is customary to set

$$z_\mu = 1^{\alpha_1}2^{\alpha_2}\cdots n^{\alpha_n}\alpha_1!\alpha_2!\cdots\alpha_n! . \qquad (1.7.5)$$

Using this and the fact that conjugacy classes are disjoint it follows that

$$\langle C_\mu, C_\lambda \rangle = \begin{cases} \dfrac{1}{n!}\dfrac{n!}{z_\mu} = \dfrac{1}{z_\mu} & \text{if } \lambda = \mu \\ 0 & \text{otherwise} . \end{cases} \qquad (1.7.6)$$

There is also a natural scalar product in the space $\Lambda^{=n}$ of symmetric polynomial homogeneous of degree n which is obtained by setting for two power basis elements p_λ, p_μ

$$\langle p_\lambda, p_\mu \rangle = \begin{cases} z_\mu & \text{if } \lambda = \mu \\ 0 & \text{otherwise} . \end{cases} \qquad (1.7.7)$$

This given, note that if we set

$$F\,C_\mu = p_\mu/z_\mu \qquad (1.7.8)$$

and use the fact that $\Lambda^{=n}$ and the center $C(S_n)$ have the same dimension, we can define an invertible linear map

$$F : C(S_n) \longleftrightarrow \Lambda^{=n}$$

which is also an isometry. In fact, note that from (1.7.6) and (1.7.7) it follows that

$$\langle FC_\mu , FC_\lambda \rangle = \tfrac{1}{z_\mu^2}\langle p_\mu , p_\lambda \rangle = \tfrac{1}{z_\mu}\chi(\lambda = \mu)] = \langle C_\mu , C_\lambda \rangle .$$

Now Frobenius introduced this map precisely for the purpose of constructing all the irreducible characters of S_n. The Frobenius result can be stated as follows.

Theorem 1.8 *The class functions*

$$\xi^\lambda = F^{-1}s_\lambda \tag{1.7.9}$$

form a complete set of the irreducible characters of S_n.

The surprising fact is that Young's and Frobenius' partition indexing of the irreducible characters turn out to be identical. This result, established by Young himself, may be simply be stated as follows:

Theorem 1.9 (A. Young)

$$F^{-1}s_\lambda = \frac{n_\lambda}{n!} \sum_{T \in \mathrm{INJ}(\lambda)} N(T)P(T) . \tag{1.7.10}$$

We should mention that to the best our knowledge there is at the moment no direct proof of this fact. Moreover, even Young's "proof" is based on a very dubious use of his quite heuristic (hair) *raising operators.* Our proof as we will later see is very indirect. Therefore it is worthwhile taking a close look at what needs to be proved. To begin we need the following very convenient form of the Frobenius map:

Proposition 1.14 *For each $\sigma \in S_n$ set*

$$\psi(\sigma) = p_{\lambda(\sigma)} \tag{1.7.11}$$

where $\lambda(\sigma)$ is the partition giving the cycle structure of σ. Then for any class function $g \in C(S_n)$ we have

$$Fg = \langle g , \psi \rangle . \tag{1.7.12}$$

Proof The proof is immediate. We need only verify (1.7.12) for the conjugacy class elements C_μ with $\mu \vdash n$. In this case the definition in (1.7.1) gives

$$F C_\mu = \langle C_\mu , \psi \rangle = \frac{1}{n!} \sum_{\sigma \in S_n} C_\mu \Big|_\sigma p_{\lambda(\sigma)} = \frac{1}{n!} p_\mu \frac{n!}{z_\mu} = \frac{p_\mu}{z_\mu}$$

which is precisely the definition in (1.7.8).   □

As a corollary we obtain

Proposition 1.15 *The equality in (1.7.10) amounts to showing that for any injective tableau T of shape λ we have*

$$s_\lambda = \frac{1}{h_\lambda} \sum_{\alpha \in R(T)} \sum_{\beta \in C(T)} \text{sign}(\beta) \, p_{\lambda(\alpha\beta)} \,. \tag{1.7.13}$$

Proof Note that (1.7.10) may also be written in the form

$$s_\lambda = \frac{n_\lambda}{n!} F \sum_{\sigma \in S_n} N(\sigma T) P(\sigma T) = \frac{n_\lambda}{n!} F \sum_{\sigma \in S_n} \sigma N(T) P(T) \sigma^{-1} \,.$$

Now, using (1.7.12) we get

$$s_\lambda = \frac{n_\lambda}{n!} \sum_{\sigma \in S_n} \langle \sigma N(T) P(TY) \sigma^{-1} \,, \, \psi \rangle$$

$$(\text{by (1.7.2) and (1.7.3)}) = \frac{n_\lambda}{n!} \sum_{\sigma \in S_n} \langle N(T) P(TY) \,, \, \sigma^{-1} \psi \sigma \rangle$$

$$(\text{since } \psi \text{ is central}) = \frac{n_\lambda}{n!} \sum_{\sigma \in S_n} \langle N(T) P(T) \,, \, \psi \rangle = n_\lambda \langle N(T) P(T) \,, \, \psi \rangle$$

$$= \frac{n_\lambda}{n!} \sum_{\alpha \in R(T)} \sum_{\beta \in C(T)} \text{sign}(\beta) \, p_{\lambda(\alpha\beta)} \,.$$

This proves (1.7.13).   □

Lecture 2
Young's Seminormal Representation, Murphy Elements, and Content Evaluations

2.1 Introduction

These are notes resulting from a course in Algebraic Combinatorics given at the University of California at San Diego in winter 2003. The lectures were based on writings of A. Young [29], R. M. Thrall [24], D. E. Rutherford [20], A. Jucys [14] and G. Murphy [17]. The material covers basic identities leading to the construction of Young's seminormal units. Murphy elements are used first as an aid to the construction of the entries in the seminormal matrices corresponding to the simple reflections, and then used to express characters and conjugacy classes. As a by-product we obtain a new way of constructing certain polynomials introduced in a more recent paper by A. Goupil et al. [13] and by Diaconis and Greene in an earlier unpublished manuscript [6]. These polynomials yield some truly remarkable formulas for the irreducible characters of the symmetric groups. More precisely, it is shown in [6] and [13] that for each partition $\gamma \vdash k$ we can construct a symmetric polynomial W_γ which evaluated at the contents of a partition $\lambda \vdash n \geq k$ yield the central character value $\omega^\lambda_{\gamma,1^{n-k}}$. The polynomials W_γ are remarkably simple when expressed in terms of the power basis. Moreover, their power basis expansion are of the form $W_\gamma = \sum_\rho c^\gamma_\rho(n) \, p_\rho$ with coefficients $c^\gamma_\rho(n)$ polynomials in n. The approach followed in [13] is quite intricate and somewhat indirect. In [6] Diaconis and Greene follow a purely combinatorial approach and derive an algorithm for constructing the polynomials W_γ. In these notes we obtain explicit closed form expressions for W_γ. Our approach is based on a method introduced by Macdonald [16] in an exercise where an explicit formula is obtained for the central character value $\omega^\lambda_{k,1^{n-k}}$.

The material is structured into five sections. We start by *Young idempotents* in which we introduce the Young's last letter order and establish some of the basic properties of Young idempotents under this order. In *Young's seminormal units* we construct the Young seminormal units and prove their orthogonality.

A. M. Garsia, Ö. Eğecioğlu, *Lectures in Algebraic Combinatorics*, Lecture Notes in Mathematics 2277, https://doi.org/10.1007/978-3-030-58373-6_2

We also construct the seminormal matrix units and compute the characters of the corresponding representations, and end this section by proving seminormality. In the section *Murphy elements* we introduce the Murphy elements and prove their commutativity. We derive the action of the class C_2 of transpositions on a seminormal tableau unit. We thus obtain in a purely combinatorial way that the central character $\chi^\lambda_{21^{n-2}} |C_2|/n_\lambda$ is $n(\lambda') - n(\lambda)$. The basic result here is that the seminormal units are simultaneously eigenfunctions of the Murphy elements. We show that the eigenvalue of m_k on the tableau unit $e(T)$ is given by the content of the cell of k in T. We also obtain in a canonical way a polynomial $P_T(m_2, \ldots, m_n)$ which gives $e(T)$. Following this, in *the seminormal matrices* we derive the entries in Young's seminormal matrices from the action of Murphy elements on the seminormal matrix units. Finally, *Murphy elements and conjugacy classes* starts by two different proofs that symmetric polynomials evaluated at the Murphy elements yield class functions. This done our main goal is to obtain explicit formulas for the symmetric polynomials that yield the characters and the classes. Tables of these polynomials are included in a few special cases.

2.2 The Young Idempotents

Now we recall some basic properties of the Young idempotents and set notation. Most of the material covered in this section is presented in full detail in the lecture notes on Young's natural representation *Alfred Young's construction of the irreducible representations of S_n*. References to these notes will be indicated by the symbol [YNR].

We shall use the French convention of depicting Ferrers diagrams as left justified rows of lattice cells of lengths decreasing from bottom to top as we did in [YNR].

Recall that a tableau T of shape $\lambda \vdash n$ is a filling of the cells of the Ferrers diagram of λ with the letters $1, 2, \ldots, n$. If the filling is row and column increasing then T is said to be *standard*. The shape of T will be simply denoted $\lambda(T)$. Given a tableau T, we let $R(T)$ and $C(T)$ respectively denote the *row group* and *column group* of T. As in [YNR] we set

$$P(T) = \sum_{\alpha \in R(T)} \alpha, \quad N(T) = \sum_{\beta \in C(T)} \text{sign}(\beta)\,\beta, \quad (2.2.1)$$

and

$$E(T) = P(T)N(T). \quad (2.2.2)$$

Note that if T has n cells and $\sigma \in S_n$ then σT denotes the tableau obtained by replacing in T the index i by σ_i for $i = 1, 2, \ldots, n$. It is easily seen that we have

$$R(\sigma T) = \sigma R(T)\sigma^{-1}, \quad C(\sigma T) = \sigma C(T)\sigma^{-1}$$

thus

$$P(\sigma T) = \sigma P(T)\sigma^{-1}, \quad N(\sigma T) = \sigma N(T)\sigma^{-1}. \tag{2.2.3}$$

In particular if T_1 and T_2 are tableaux of the same shape and σ_{T_1,T_2} is the permutation that sends T_2 into T_1 we have the identities

$$\sigma_{T_1,T_2} P(T_2) = P(T_1)\sigma_{T_1,T_2},$$
$$\sigma_{T_1,T_2} N(T_2) = N(T_1)\sigma_{T_1,T_2}, \tag{2.2.4}$$
$$\sigma_{T_1,T_2} E(T_2) = E(T_1)\sigma_{T_1,T_2}.$$

We have the following basic fact

Proposition 2.1 *If T_1 and T_2 are two tableaux with n cells, then we have*

$$N(T_2)P(T_1) \neq 0 \quad or \quad P(T_1)N(T_2) \neq 0 \tag{2.2.5}$$

only if

$$\lambda(T_2) \geq \lambda(T_1) \quad (in\ dominance). \tag{2.2.6}$$

In particular

$$\lambda(T_2) < \lambda(T_1) \implies N(T_2)P(T_1) = P(T_1)N(T_2) = 0.$$

Proof Construct the diagram $T_1 \wedge T_2$ by placing in the lattice cell (i, j) the intersection of row i of T_1 with column j of T_2. Note that if one of these intersections contains two elements r and s then $R(T_1)$ and $C(T_2)$ would have the transposition (r, s) in common, but then we would have, for instance

$$N(T_2)P(T_1) = N(T_2)(r, s)P(T_1) = -N(T_2)P(T_1)$$

in plain contradiction with (2.2.5). Thus (2.2.5) forces the cells of $T_1 \wedge T_2$ to contain at most one element. Note that if we lower these elements down their columns by eliminating the empty cells we will obtain a tableau T_3 of shape $\lambda(T_2)$, thus the number of cells in the first i rows of T_3 is given by

$$\lambda_1(T_2) + \lambda_2(T_2) + \cdots + \lambda_i(T_2).$$

But all the elements of the first i rows of T_1 are in the first i rows of $T_1 \wedge T_2$ and a fortiori must all be in the first i rows of T_3. This gives the inequality

$$\lambda_1(T_2) + \lambda_2(T_2) + \cdots + \lambda_i(T_2) \geq \lambda_1(T_1) + \lambda_2(T_1) + \cdots + \lambda_i(T_1).$$

Since this must hold true for all i we necessarily have (2.2.6) as asserted. □

It will be convenient to denote the group algebra of S_n by $\mathcal{A}[S_n]$. This given, Proposition 2.1 implies the following fundamental fact.

Theorem 2.1 *If the tableaux T_1, T_2 have both n cells, then for any element $f \in \mathcal{A}[S_n]$ we have*

$$P(T_1) f N(T_2) \neq 0 \quad or \quad N(T_2) f P(T_1) \neq 0 \implies \lambda(T_2) \geq \lambda(T_1) . \qquad (2.2.7)$$

In particular

$$\lambda(T_1) \neq \lambda(T_2) \implies E(T_1) f E(T_2) = 0 \quad (for\ all\ f \in \mathcal{A}[S_n]) . \qquad (2.2.8)$$

Proof Note that we may write (using (2.2.3))

$$P(T_1) f N(T_2) = \sum_{\sigma \in S_n} f(\sigma) P(T_1) \sigma N(T_2) = \sum_{\sigma \in S_n} f(\sigma) P(T_1) N(\sigma T_2) \sigma^{-1} .$$

Thus the nonvanishing of the left hand side forces the nonvanishing at least one term in the sum in the right hand side. But then (2.2.7) follows from (2.2.6). Of course the same will hold true if $N(T_2) f P(T_1) \neq 0$. But now note that

$$P(T_1) N(T_1) f P(T_2) N(T_2) \neq 0 \qquad (2.2.9)$$

forces $N(T_1) f P(T_2) \neq 0$ yielding

$$\lambda(T_1) \geq \lambda(T_2) .$$

On the other hand, (by (2.2.7)), (2.2.9) itself forces

$$\lambda(T_2) \geq \lambda(T_1) .$$

Thus only the equality $\lambda(T_1) = \lambda(T_2)$ is compatible with the nonvanishing of $E(T_1) f E(T_2)$. This proves (2.2.8) and completes the proof. $\qquad \square$

We have the following fundamental identity from [YNR].

Proposition 2.2 (Von Neuman Sandwich Lemma) *For any element $f \in \mathcal{A}(S_n)$ and any tableau T*

$$P(T) f N(T) = c_T(f) P(T) N(T) , \qquad (2.2.10)$$

with

$$c_T(f) = f N(T) P(T) |_\epsilon ,$$

where ϵ denotes the identity permutation.

This result has the following important corollary.

Theorem 2.2 *For any tableau T, the group algebra element $E(T)$ is idempotent. More precisely there is a nonvanishing constant $h(T)$ depending only on the shape of T such that*

$$E(T)E(T) = h(T) E(T). \qquad (2.2.11)$$

Proof Using (2.2.10) with $f = N(T)P(T)$, the identity in (2.2.10) gives

$$E(T)E(T) = h(T)E(T)$$

with

$$h(T) = N(T)P(T)N(T)P(T)\big|_\epsilon .$$

Note that if $h(T)$ were to vanish then $E(T)$ would be nilpotent. Now a cute argument shows that that the coefficient of the identity of any nilpotent element of a group algebra necessarily vanishes. Now this immediately leads to a contradiction since we can easily see that

$$E(T)\big|_\epsilon = \sum_{\alpha \in R(T)} \sum_{\beta \in C(T)} \operatorname{sign}(\beta)\,\alpha\beta \big|_\epsilon = 1.$$

In fact the identity $\alpha\beta = \epsilon$ holds only when $\alpha = \beta = \epsilon$. We shall show later that if T has shape λ then

$$h(T) = \frac{n!}{n_\lambda}$$

where n_λ denotes the number of standard tableaux of shape λ. But for the moment it will suffice to show that $h(T)$ depends only on $\lambda(T)$. Now this follows immediately from the identities

$$h(T) = N(T)P(T)N(T)P(T)\big|_\epsilon$$

$$= \frac{1}{n!} \sum_{\sigma \in S_n} \sigma\, N(T)P(T)N(T)P(T)\,\sigma^{-1}\big|_\epsilon$$

$$= \frac{1}{n!} \sum_{\sigma \in S_n} N(\sigma T)P(\sigma T)N(\sigma T)P(\sigma T)\big|_\epsilon$$

$$= \frac{1}{n!} \sum_{\lambda(T')=\lambda(T)} N(T')P(T')N(T')P(T')\big|_\epsilon .$$

$\square$

This given, here and after we will use the symbol h_λ to denote $h(T)$ for any tableau T of shape λ.

In the construction of Young's seminormal representation we will make systematic use of Young's *last letter order* as in [YNR]. Given two standard tableaux T_1, T_2 of the same shape we shall say that k is the *last letter of disagreement* between T_1 and T_2 if the letters $k+1, k+2, \ldots, n$ occupy exactly the same positions in T_1 and T_2, but k does not. This given we shall say that

"T_1 precedes T_2 in the Young's last letter order"

and write

$$T_1 <_{LL} T_2 \tag{2.2.12}$$

if and only if the last letter of disagreement is *higher* in T_1 than in T_2.

If T is a standard tableau we shall denote by $T(k)$ the tableau obtained by removing from T the letters larger than k together with their cells. Note that if $\lambda(T_1) = \lambda(T_2)$ and $k+1$ is the last letter of disagreement between T_1 and T_2 then $\lambda\big(T_1(k+1)\big) = \lambda\big(T_2(k+1)\big)$ and (2.2.12) holds true if and only if

$$\lambda(T_1(k)) > \lambda(T_2(k)) \qquad \text{(in dominance)} . \tag{2.2.13}$$

For $f \in \mathcal{A}(S_n)$ let us set

$$\downarrow f = \sum_{\sigma \in S_n} f(\sigma)\sigma^{-1} \tag{2.2.14}$$

We shall refer to this operation as *flipping* f. There are interesting identities connected with the operation of passing from T to $T(k)$. They may be stated as follows.

Proposition 2.3 *For any tableau T of shape $\lambda \vdash n$ and $k \leq n$ we have two elements $p_k(T)$ and $n_k(T)$ such that*

$$\begin{aligned}
&(a)\ \ p_k(T)P(T(k)) = P(T) = P(T(k))\downarrow p_k(T), \\
&(b)\ \ n_k(T)N(T(k)) = N(T) = N(T(k))\downarrow n_k(T).
\end{aligned} \tag{2.2.15}$$

Proof Note that the second equality in (2.2.15) (a) follows by flipping both sides of the first equality. The same holds true for (2.2.15 (b)). So in each case we need only show the first equality. Moreover since we can pass from T to $T(k)$ by successive removals of the largest letter, we can see that we need only show our equalities in the case $k = n - 1$. To this end let us suppose that the row of T that contains the letter n consists of the letters

$$a_1, a_2, \ldots, a_r, n .$$

In this case denoting by

$$(a_1, a_2, \ldots, a_r, n) \quad \text{and} \quad (a_1, a_2, \ldots, a_r)$$

the formal sums of all the permutations of S_n that leave fixed the elements of the complements of $\{a_1, a_2, \ldots, a_r, n\}$ and $\{a_1, a_2, \ldots, a_r\}$ respectively, we have, by left coset decomposition

$$(a_1, a_2, \ldots, a_r, n) = \big(\epsilon + (a_1, n) + (a_2, n) + \cdots + (a_r, n)\big)(a_1, a_2, \ldots, a_r).$$

Multiplying this identity by the contributions to $P(T)$ coming from the other rows of T yields

$$P(T) = \big(\epsilon + (a_1, n) + (a_2, n) + \cdots + (a_r, n)\big) P(T(n-1))$$

which is precisely the first equality in (2.2.15) (a) for $k = n - 1$.

For (2.2.15) (b) we can proceed in a similar way. Indeed let us suppose that the column of T that contains the letter n consists of the letters

$$b_1, b_2, \ldots, b_s, n.$$

Denoting by

$$(b_1, b_2, \ldots, b_s, n)' \quad \text{and} \quad (b_1, b_2, \ldots, a_s)'$$

the formal sums of all the signed permutations of S_n that leave fixed the elements of the complements of $\{b_1, b_2, \ldots, b_s, n\}$ and $\{b_1, b_2, \ldots, b_s\}$ respectively, we have, by left coset decomposition

$$(b_1, b_2, \ldots, b_s, n)' = \big(\epsilon - (b_1, n) - (b_2, n) - \cdots - (b_s, n)\big)(b_1, b_2, \ldots, b_r)'.$$

Multiplying this identity by the contributions to $N(T)$ coming from the other columns of T yields

$$N(T) = \big(\epsilon - (b_1, n) - (b_2, n) - \cdots - (b_r, n)\big) N(T(n-1))$$

which is precisely the first equality in (2.2.15) (b) for $k = n - 1$. This completes our proof. □

The following result will play a crucial role here.

Theorem 2.3 *For two standard tableau T_1 and T_2 of the same shape we have*

$$N(T_1)P(T_2) \neq 0 \quad \Longrightarrow \quad T_1 <_{LL} T_2. \qquad (2.2.16)$$

Proof Let $k + 1$ be the last letter of disagreement between T_1 and T_2. Using Proposition 2.3 we derive the factorization

$$N(T_1)P(T_2) = n_k(T_1) \, N(T_1(k))P(T_2(k)) \downarrow p_k(T_2) .$$

Thus

$$N(T_1)P(T_2) \neq 0 \quad \Longrightarrow \quad N(T_1(k))P(T_2(k)) \neq 0$$

and Proposition 2.1 gives

$$\lambda((T_1(k)) > \lambda(T_2(k)) .$$

However we have seen that this holds true if and only if

$$T_1 <_{LL} T_2 .$$

This completes our argument.   □

2.3   Young's Seminormal Units

As we have shown in [YNR] for the case of the symmetric group, the group algebra $\mathcal{A}(G)$ of a finite group G has a basis

$$\{\{e_{ij}^{\lambda}\}_{i,j=1}^{n_\lambda}\}_{\lambda \in \Lambda} \tag{2.3.1}$$

with the property that

$$e_{ij}^{\lambda} e_{rs}^{\mu} = \begin{cases} 0 & \text{if } \lambda \neq \mu , \\ 0 & \text{if } \lambda = \mu , \text{ but } j \neq r \\ e_{is}^{\mu} & \text{if } \lambda = \mu \text{ and } j = r . \end{cases} \tag{2.3.2}$$

Moreover, we have the identities

$$e_{ij}^{\lambda} \big|_{\epsilon} = \begin{cases} 0 & \text{if } i \neq j , \\ \frac{1}{h_\lambda} & \text{if } i = j . \end{cases} \qquad (\text{with } h_\lambda = |G|/n_\lambda) \tag{2.3.3}$$

From these identities it follows that we have the expansions

$$f = \sum_{\lambda \in \Lambda} h_\lambda \sum_{i,j=1}^{n_\lambda} f e_{ji}^{\lambda} \big|_{\epsilon} e_{ij}^{\lambda} \qquad (\text{for all } f \in \mathcal{A}(G)). \tag{2.3.4}$$

This basis allows us to construct a complete set of representatives of irreducible representations of G. These are simply given by the collection $\{A^\lambda\}_{\lambda \in \Lambda}$ obtained by setting

$$A^\lambda(\sigma) = \|a_{ij}^\lambda(\sigma)\|_{i,j=1}^{n_\lambda} \tag{2.3.5}$$

with

$$a_{ij}^\lambda(\sigma) = h_\lambda e_{ji}^\lambda\big|_{\sigma^{-1}} \qquad \text{(for all } \sigma \in G\text{)}. \tag{2.3.6}$$

In fact, from (2.3.4) and (2.3.6) we derive the expansion

$$\sigma = \sum_{\lambda \in \Lambda} \sum_{i,j=1}^{n_\lambda} a_{ij}^\lambda(\sigma) \, e_{ij}^\lambda \tag{2.3.7}$$

and the relations in (2.3.3) immediately yield that

$$\sigma e_{r,s}^\mu = \sum_{i=1}^{n_\mu} e_{i,s}^\mu \, a_{i,r}^\mu(\sigma) \tag{2.3.8}$$

or in matrix form

$$\sigma \big(e_{i,s}^\mu\big)_{1 \le i \le n_\mu} = \big(e_{i,s}^\mu\big)_{1 \le i \le n_\mu} A^\mu(\sigma). \tag{2.3.9}$$

Now we see from (2.3.6) that the representation A^λ is orthogonal (that is $a_{ij}(\sigma^{-1}) = a_{ji}(\sigma)$) if and only if we have

$$e_{ji}^\lambda(\sigma) = e_{ij}^\lambda(\sigma^{-1}) \qquad \text{(for all } \sigma \in G\text{)} \tag{2.3.10}$$

and this, in compact form may be simply be rewritten as

$$\downarrow e_{ij}^\lambda = e_{ji}^\lambda. \tag{2.3.11}$$

When Young set himself the task of finding a complete set of irreducible orthogonal representations of S_n, his point of departure (see [29, QSA VI]) was the construction of units e_{ij}^λ satisfying conditions (2.3.2) and (2.3.3) together with the additional condition

$$\downarrow e_{ij}^\lambda = \frac{d_i^\lambda}{d_j^\lambda} e_{ji}^\lambda. \tag{2.3.12}$$

These constructs have come to be referred to as *Young's seminormal units*. It is easily seen that setting

$$f_{ij}^{\lambda} = \sqrt{\frac{d_j^{\lambda}}{d_i^{\lambda}}} \, e_{ij}^{\lambda} \, , \tag{2.3.13}$$

the orthogonality condition in (2.3.11) will be satisfied by the f_{ij}^{λ} and the desired orthogonal representations can then be readily obtained.

Surprisingly Young's seminormal units can be written down with a minimal amount of additional notation from the material which is already in our possession. To begin, we let

$$T_1^{\lambda}, T_2^{\lambda}, \ldots, T_{n_{\lambda}}^{\lambda}$$

denote the standard tableaux of shape λ in Young's last letter order. Moreover the permutation that sends T_j^{λ} onto T_i^{λ} will be denoted σ_{ij}^{λ}. In symbols

$$\sigma_{ij}^{\lambda} \, T_j^{\lambda} = T_i^{\lambda} \, . \tag{2.3.14}$$

Finally, for a tableau T with n cells, it will be convenient, to use the abbreviations

$$T(n-1) = \overline{T}, \quad T(n-2) = \overline{\overline{T}} \, .$$

This given, Young's seminormal units are simply given by the formulas

$$e_{ij}^{\lambda} = e(\overline{T}_i^{\lambda}) \, \frac{P(T_i^{\lambda}) \, \sigma_{ij}^{\lambda} \, N(T_j^{\lambda})}{h_{\lambda}} \, e(\overline{T}_j^{\lambda}) \, , \tag{2.3.15}$$

where for a given standard tableau T of shape λ the group algebra element $e(T)$ is recursively defined by setting

$$e(T) = \begin{cases} \epsilon & \text{if } T = [1], \\ e(\overline{T}) \frac{P(T)N(T)}{h_{\lambda}} e(\overline{T}) & \text{otherwise.} \end{cases} \tag{2.3.16}$$

The remainder of this section is dedicated to proving that these units satisfy the identities in (2.3.2), (2.3.3), and (2.3.12). But before we can carry this out we need to derive a few additional properties.

Our basic tool is provided by the following.

Proposition 2.4 *For any standard tableau T we have*

$$(a)\ \ E(T)e(\overline{T})E(T)\ =\ h(T)\,E(T)$$
$$(b)\ \ E(T)e(T)E(T)\ =\ h(T)\,E(T) \qquad (2.3.17)$$
$$(c)\ \ e(T)E(T)e(T)\ =\ h(T)\,e(T)$$

where we set $h(T) = h_\lambda$ *when* $\lambda(T) = \lambda$.

Proof To prove this we need two auxiliary identities which are of independent interest.

Lemma 2.1

$$(a)\ \ e(T)P(T)N(T)\big|_\epsilon\ =\ 1\,,$$
$$(b)\ \ e(\overline{T})N(T)P(T)\big|_\epsilon\ =\ 1\,. \qquad (2.3.18)$$

Proof We proceed by induction on the size of T. So assume that (2.3.18) (a) is true for any tableau with $n-1$ cells and let T be a tableau with n cells. Note that the induction hypothesis then implies that

$$e(\overline{T})P(\overline{T})N(\overline{T})\big|_\epsilon\ =\ 1\,. \qquad (2.3.19)$$

Now let the elements of T that are in the same row and column as n be given as in the proof of Proposition 2.3. We may then write

$$\sigma\,P(T)N(T)\big|_\epsilon \qquad (2.3.20)$$
$$= P(\overline{T})\big(\epsilon + (a_1,n) + \cdots + (a_r,n)\big)\big(\epsilon - (b_1,n) - \cdots - (b_s,n)\big)N(\overline{T})\big|_{\sigma^{-1}}.$$

Note that for any pair i, j, the product of the two cycles $(a_i,n)(b_j,n)$ is the 3-cycle (a_i,n,b_j) which is not in S_{n-1}, and since all of the permutations in $P(\overline{T})$ or $N(\overline{T})$ are in S_{n-1}, the only terms in (2.3.20) that will yield permutations in S_{n-1} are those produced by the identity term ϵ. Thus we necessarily have

$$\sigma\,P(T)N(T)\big|_\epsilon = P(\overline{T})N(\overline{T})\big|_{\sigma^{-1}} = \sigma\,P(\overline{T})N(\overline{T})\big|_\epsilon \quad \text{(for all } \sigma \in S_{n-1}).$$

Multiplying this relation by $e(\overline{T})\big|_\sigma$ and summing over $\sigma \in S_{n-1}$ we get

$$e(\overline{T})\,P(T)N(T)\big|_\epsilon\ =\ e(\overline{T})\,P(\overline{T})N(\overline{T})\big|_\epsilon$$

and (2.3.19) gives

$$e(\overline{T})\,P(T)N(T)\big|_\epsilon\ =\ 1\,,$$

proving (2.3.18) (b). Now (2.3.16) gives

$$h(T)e(T)P(T)N(T)\big|_\epsilon = e(\overline{T})P(T)N(T)e(\overline{T})P(T)N(T)\big|_\epsilon$$

$$\text{(by Proposition 2.2)} = e(\overline{T})\Big(N(T)e(\overline{T})P(T)N(T)P(T)\big|_\epsilon\Big)P(T)N(T)\big|_\epsilon$$

$$\text{(by (2.3.18) (b)} = \Big(e(\overline{T})P(T)N(T)P(T)N(T)\big|_\epsilon\Big)$$

$$\text{(by (2.2.11))} = h(T)\,e(\overline{T})P(T)N(T)\big|_\epsilon$$

$$\text{(by (2.3.18) (b)} = h(T)\,.$$

This proves

$$e(T)P(T)N(T)\big|_\epsilon = 1$$

and completes the induction.   □

After the proof of Lemma 2.1 we continue with the proof of Proposition 2.4. We have

$$E(T)e(\overline{T})E(T) = P(T)N(T)e(\overline{T})P(T)N(T)$$

$$\text{(by Proposition 2.2)} = N(T)e(\overline{T})P(T)N(T)P(T)\big|_\epsilon P(T)N(T)$$

$$= e(\overline{T})P(T)N(T)P(T)N(T)\big|_\epsilon P(T)N(T)$$

$$\text{(by (2.2.11))} = h(T)e(\overline{T})P(T)N(T)\big|_\epsilon P(T)N(T)$$

$$\text{(by (2.3.18) (b))} = h(T)\,P(T)N(T) = h(T)E(T)\,.$$

This proves (2.3.17) (a).

Next we have

$$E(T)e(T)E(T) = P(T)N(T)e(T)P(T)N(T)$$

$$\text{(by Proposition 2.2)} = N(T)e(T)P(T)N(T)P(T)\big|_\epsilon P(T)N(T)$$

$$= e(T)P(T)N(T)P(T)N(T)\big|_\epsilon P(T)N(T)$$

$$\text{(by (2.2.11))} = h(T)e(T)P(T)N(T)\big|_\epsilon P(T)N(T)$$

$$\text{(by (2.3.18) (a))} = h(T)\,P(T)N(T) = h(T)E(T)\,.$$

This proves (2.3.17) (b).

Finally from the definition in (2.3.16) we get

$$e(T)E(T)e(T) = e(\overline{T})\frac{E(T)}{h(T)}e(\overline{T})E(T)e(T)$$

$$\text{(by (2.3.17) (a))} = e(\overline{T})E(T)e(T)$$

$$\text{(by (2.3.16))} = e(\overline{T})E(T)e(\overline{T})\frac{E(T)}{h(T)}e(\overline{T})$$

$$\text{(by (2.3.17) (a))} = e(\overline{T})E(T)e(\overline{T})$$

$$\text{(by (2.3.16))} = h(T)e(T).$$

This proves (2.3.17) (c) and completes the proof of Proposition 2.4. □

We are now ready to establish the basic properties of Young's seminormal units. We begin with a result which is an immediate consequence of the definition.

Theorem 2.4 *For any standard tableau T the group algebra element $e(T)$ is idempotent.*

Proof For $T = [1]$ this is true by definition, we can thus proceed by induction on the number of cells of T. Now we have

$$e(T)e(T) = e(\overline{T})\frac{E(T)}{h(T)}e(\overline{T})e(\overline{T})\frac{E(T)}{h(T)}e(\overline{T})$$

$$\text{(by induction)} = e(\overline{T})\frac{E(T)}{h(T)}e(\overline{T})\frac{E(T)}{h(T)}e(\overline{T})$$

$$\text{(by (2.3.17) (a))} = e(\overline{T})\frac{E(T)}{h(T)}e(\overline{T}) = e(T).$$

These idempotents are orthogonal, more precisely we have

Theorem 2.5 *If T_1 and T_2 are standard tableaux with n cells then*

$$e(T_1)e(T_2) = \begin{cases} 0 & \text{if } \lambda(T_1) \neq \lambda(T_2) \\ 0 & \text{if } \lambda(T_1) = \lambda(T_2) \text{ but } T_1 \neq T_2 \\ e(T_1) & \text{if } T_1 = T_2. \end{cases} \quad (2.3.21)$$

Proof Since by definition

$$e(T_1)e(T_2) = e(\overline{T}_1)\frac{E(T_1)}{h(T_1)}e(\overline{T}_1)e(\overline{T}_2)\frac{E(T_2)}{h(T_2)}e(\overline{T}_2), \quad (2.3.22)$$

the first equality is an immediate consequence of (2.2.8). The last equality is a restatement of Theorem 2.4. We are left to prove the second equality. Since for $n = 1, 2$ there is nothing to prove, we shall proceed by induction on n and assume its validity for $n - 1$. This given we have the following sequence of implications

$$e(T_1)\,e(T_2) \neq 0$$

(by (2.3.22)) $\Downarrow$

$$e(\overline{T}_1)\,e(\overline{T}_2) \neq 0$$

(by induction) $\Downarrow$

$$\overline{T}_1 = \overline{T}_2$$

(when $\lambda(T_1) = \lambda(T_2)$) $\Downarrow$

$$T_1 = T_2$$

which prove our assertion. $\square$

Theorem 2.6 *Young's seminormal units satisfy the identities*

$$
e_{ij}^{\lambda} e_{rs}^{\mu} =
\begin{cases}
0 & \text{if } \lambda \neq \mu, \\
0 & \text{if } \lambda = \mu, \text{ but } j \neq r \\
e_{is}^{\mu} & \text{if } \lambda = \mu \text{ and } j = r.
\end{cases}
\tag{2.3.23}
$$

Proof From the definition in (2.3.15) we derive that

$$(a)\ e_{ij}^{\lambda} = e(\overline{T}_i^{\lambda})\,\sigma_{ij}^{\lambda}\,\frac{E(T_j^{\lambda})}{h_\lambda}\,e(\overline{T}_j^{\lambda}), \quad (b)\ e_{rs}^{\mu} = e(\overline{T}_r^{\mu})\,\frac{E(T_r^{\mu})}{h_\mu}\,\sigma_{rs}^{\mu}\,e(\overline{T}_s^{\mu}).$$

$$\tag{2.3.24}$$

Thus the first equality in (2.3.23) is an immediate consequence of (2.2.8). In the case $\mu = \lambda$, (2.3.24) (a) and (b) give

$$e_{ij}^{\lambda}\,e_{rs}^{\lambda} = e(\overline{T}_i^{\lambda})\,\sigma_{ij}^{\lambda}\,\frac{E(T_j^{\lambda})}{h_\lambda}\,e(\overline{T}_j^{\lambda})e(\overline{T}_r^{\lambda})\,\frac{E(T_r^{\mu})}{h_\lambda}\,\sigma_{rs}^{\lambda}\,e(\overline{T}_s^{\lambda}).\tag{2.3.25}$$

Now if $r \neq j$, Theorem 2.5 gives $e(\overline{T}_j^{\lambda})e(\overline{T}_r^{\lambda}) = 0$, proving the second case of (2.3.23). Finally for $\lambda = \mu$ and $j = r$, (2.3.25) becomes

$$e_{ij}^{\lambda}\,e_{js}^{\lambda'} = e(\overline{T}_i^{\lambda})\,\sigma_{ij}^{\lambda}\,\frac{E(T_j^{\lambda})}{h_\lambda}\,e(\overline{T}_j^{\lambda})e(\overline{T}_j^{\lambda})\,\frac{E(T_j^{\mu})}{h_\lambda}\,\sigma_{js}^{\lambda}\,e(\overline{T}_s^{\lambda}),$$

$$(\text{by Theorem 2.4}) = e(\overline{T}_i^{\lambda})\,\sigma_{ij}^{\lambda}\,\frac{E(T_j^{\lambda})}{h_\lambda}\,e(\overline{T}_j^{\lambda})\,\frac{E(T_j^{\mu})}{h_\lambda}\,\sigma_{js}^{\lambda}\,e(\overline{T}_s^{\lambda}),$$

$$\left(\text{by (2.3.17) (a)}\right) = e(\overline{T}{}_i^\lambda)\,\sigma_{ij}^\lambda\,\frac{E(T_j^\lambda)}{h_\lambda}\,\sigma_{js}^\lambda\,e(\overline{T}{}_s^\lambda),$$

$$\left(\text{by (2.2.4)}\right) = e(\overline{T}{}_i^\lambda)\,\frac{P(T_i^\lambda)\,\sigma_{is}^\lambda\,N(T_s^\lambda)}{h_\lambda}\,e(\overline{T}{}_s^\lambda) = e_{is}^\lambda. \tag{2.3.26}$$

This proves the last equality in (2.3.23) and completes our argument. □

Note that we also have

Theorem 2.7

$$e_{ij}^\lambda\big|_\epsilon = \begin{cases} 0 & \text{if } i \neq j, \\ 1/h_\lambda & \text{if } i = j. \end{cases} \tag{2.3.27}$$

Proof The definition in (2.3.15) gives

$$\begin{aligned}
e_{ij}^\lambda\big|_\epsilon &= e(\overline{T}{}_i^\lambda)\,\frac{P(T_i^\lambda)\,\sigma_{ij}^\lambda\,N(T_j^\lambda)}{h_\lambda}\,e(\overline{T}{}_j^\lambda)\bigg|_\epsilon \\
&= e(\overline{T}{}_j^\lambda)\,e(\overline{T}{}_i^\lambda)\,\frac{P(T_i^\lambda)\,\sigma_{ij}^\lambda\,N(T_j^\lambda)}{h_\lambda}\bigg|_\epsilon
\end{aligned}$$

and the first case of (2.3.27) follows from Theorem 2.5. But for $i = j$ this becomes

$$e_{ii}^\lambda\big|_\epsilon = e(\overline{T}{}_i^\lambda)\,e(\overline{T}{}_i^\lambda)\,\frac{P(T_i^\lambda)\,N(T_i^\lambda)}{h_\lambda}\bigg|_\epsilon$$

$$\left(\text{by Theorem 2.4}\right) = \frac{1}{h_\lambda}\,e(\overline{T}{}_i^\lambda)\,P(T_i^\lambda)\,N(T_i^\lambda)\bigg|_\epsilon$$

$$\left(\text{by (2.3.18) (b)}\right) = \frac{1}{h_\lambda}.$$

This completes our proof. □

From the identities in (2.3.23) and formula (2.3.9), it follows that, for $\lambda \vdash n$, the character of the representation resulting from the action of S_n on a linear span

$$\mathcal{L}[e_{1s}^\lambda,\, e_{2s}^\lambda,\, \ldots,\, e_{n_\lambda s}^\lambda]$$

is the same as the character of the action of S_n on the left ideal $\mathcal{A}(S_n)e(T_s^\lambda)$. From this observation we derive the following important fact.

Proposition 2.5 *The character of the action of S_n on $\mathcal{L}\left[e_{1s}^\lambda, e_{2s}^\lambda, \dots, e_{n_\lambda s}^\lambda\right]$ depends only on λ and it is given by the formula*

$$\chi^\lambda = \sum_{T \in T(\lambda)} \frac{P(T)N(T)}{h_\lambda}$$

where $T \in T(\lambda)$ is to indicate that the sum is over all tableaux of shape λ.

Proof A basic fact from representation theory that we shall make use of is the following: If $\mathbf{e}$ is an idempotent of the group algebra $\mathcal{A}(G)$ then the character χ^e of the left multiplication action of G on the ideal $\mathcal{A}(G)\mathbf{e}$ is given by the formula

$$\chi^e = \sum_{\sigma \in G} \sigma\, \mathbf{e}\, \sigma^{-1}.$$

Taking $T = T_s^\lambda$, from this formula we obtain that

$$\chi^\lambda = \sum_{\sigma \in G} \sigma\, e(T)\sigma^{-1} = \sum_{\sigma \in S_n} \sigma\, e(\overline{T})\, \frac{P(T)N(T)}{h_\lambda}\, e(\overline{T})\sigma^{-1}$$

$$\text{(Circular rearrangements are OK)} = \sum_{\sigma \in S_n} \sigma\, e(\overline{T})e(\overline{T})\, \frac{P(T)N(T)}{h_\lambda}\, \sigma^{-1}$$

$$\text{(by Theorem 2.4)} = \sum_{\sigma \in S_n} \sigma\, e(\overline{T})\, \frac{P(T)N(T)}{h_\lambda}\, \sigma^{-1}$$

$$= \sum_{\sigma \in S_n} \sigma\, \frac{N(T)\,e(\overline{T})\,P(T)}{h_\lambda}\, \sigma^{-1}$$

$$\text{(by Von Neumann lemma)} = \left(e(\overline{T})P(T)N(T)|_\epsilon\right) \sum_{\sigma \in S_n} \sigma\, \frac{N(T)P(T)}{h_\lambda}\, \sigma^{-1}$$

and (2.3.18) (b) gives (by (2.2.3))

$$\chi^\lambda = \sum_{\sigma \in S_n} \frac{N(\sigma T)P(\sigma T)}{h_\lambda}.$$

This proves the Theorem since σT describes all tableaux of shape λ, as σ varies in S_n.   □

Remark 2.1 It is easily derived from the identities in (2.3.23) that the set $\left\{\{e_{ij}^\lambda\}_{i,j=1}^{n_\lambda}\right\}_{\lambda \vdash n}$ is independent. Furthermore, as proved in [YNR], the numbers of standard tableaux n_λ satisfy the identity

$$\sum_{\lambda \vdash n} n_\lambda^2 = n!.$$

Thus $\{\{e^\lambda_{ij}\}^{n_\lambda}_{i,j=1}\}_{\lambda \vdash n}$ is an independent subset of $\mathcal{A}(S_n)$ of cardinality equal to the dimension of $\mathcal{A}(S_n)$. So it must be a basis. From this it is easy to derive that the expansion of any element $f \in \mathcal{A}(S_n)$ in terms of the Young's seminormal units is given by formula (2.3.4).

We conclude this section with two important applications of this formula, beginning with the following beautiful identity of Alfred Young.

Theorem 2.8 *For any standard tableau T*

$$e(\overline{T}) = \sum_S \chi(\overline{S} = \overline{T})\, e(S) \tag{2.3.28}$$

where the sum is over all standard tableaux S yielding $\overline{T}$ upon removal of n.

Proof Using formula (2.3.4) for $f = e(\overline{T})$ and Young's units we get, (if T has n cells)

$$e(\overline{T}) = \sum_{\lambda \vdash n} h_\lambda \sum_{i,j=1}^{n_\lambda} e(\overline{T})\, e^\lambda_{ji}\Big|_\epsilon e^\lambda_{ij}\, . \tag{2.3.29}$$

Now we have

$$\begin{aligned}
e(\overline{T})\, e^\lambda_{ji}\Big|_\epsilon &= e(\overline{T}) e(\overline{T}^\lambda_j) \frac{P(T^\lambda_j)\sigma^\lambda_{ji} N(T^\lambda_i)}{h\lambda}\, e(\overline{T}^\lambda_i)\Big|_\epsilon \\
&= e(\overline{T}^\lambda_i) e(\overline{T}) e(\overline{T}^\lambda_j) \frac{P(T^\lambda_j)\sigma^\lambda_{ji} N(T^\lambda_i)}{h\lambda}\Big|_\epsilon\, ,
\end{aligned} \tag{2.3.30}$$

and we immediately derive from Theorem 2.5 that this term does not vanish only if

$$\overline{T}^\lambda_i = \overline{T} = \overline{T}^\lambda_j\, . \tag{2.3.31}$$

In particular this forces $i = j$, and (2.3.30) becomes

$$\begin{aligned}
e(\overline{T})\, e^\lambda_{ii}\Big|_\epsilon &= e(\overline{T}^\lambda_i) e(\overline{T}) e(\overline{T}^\lambda_i) \frac{P(T^\lambda_i) N(T^\lambda_i)}{h_\lambda}\Big|_\epsilon \\
\text{(by Theorem 2.4)} &= e(\overline{T}^\lambda_i) \frac{P(T^\lambda_j) N(T^\lambda_i)}{h_\lambda}\Big|_\epsilon \\
\text{(by (2.3.18) (b))} &= \frac{1}{h_\lambda}\, .
\end{aligned} \tag{2.3.32}$$

Using (2.3.31) and (2.3.32) reduces (2.3.29) to

$$e(\overline{T}) = \sum_{\lambda} \sum_{i=1}^{n_\lambda} \chi(\overline{T}_i^\lambda = \overline{T}) \, e_{ii}^\lambda \,. \tag{2.3.33}$$

Since

$$e_{ii}^\lambda = e(\overline{T}_i^\lambda) \frac{P(T_i^\lambda) N(T_i^\lambda)}{h_\lambda} \, e(\overline{T}_i^\lambda) = e(T_i^\lambda)$$

formula (2.3.33) is only another way of writing (2.3.28).   □

A immediate corollary of Theorem 2.8 is the identification of the constant h_λ.

Proposition 2.6

$$h_\lambda = \frac{n!}{n_\lambda} \,. \tag{2.3.34}$$

Proof Equating coefficients of the identity on both sides of (2.3.28), and using (2.3.27) we get

$$\frac{1}{h(\overline{T})} = \sum_S \chi(\overline{S} = \overline{T}) \frac{1}{h(S)} \,.$$

Assuming that T has n cells, and that $\lambda(\overline{T}) = \mu$ we may rewrite this identity in the form

$$\frac{1}{h_\mu} = \sum_\lambda \chi(\mu \rightarrow \lambda) \frac{1}{h_\lambda} \,, \tag{2.3.35}$$

where the symbol $\mu \rightarrow \lambda$ is to express that λ is obtained by adding a cell to μ. Multiplying both sides of (2.3.34) by $n!$ and setting

$$k_\mu = \frac{(n-1)!}{h_\mu} \,, \quad k_\lambda = \frac{n!}{h_\lambda}$$

converts (2.3.35) into the identity

$$n \, k_\mu = \sum_\lambda \chi(\mu \rightarrow \lambda) k_\lambda$$

which is precisely the recursion satisfied by the number of standard tableaux. From this it easily follows that for every λ we must have

$$k_\lambda = n_\lambda$$

as desired.   □

The seminormality of the Young units is based on the following remarkable fact.

Theorem 2.9 *For any standard tableaux T we have*

$$\downarrow e(T) = e(T). \tag{2.3.36}$$

Proof Since (2.3.36) is obviously true for $T = [1]$ we can proceed by induction on the number of cells of T. We have

$$\downarrow e(T) = e(\overline{T})\frac{N(T)P(T)}{h(T)}e(\overline{T}).$$

Using (2.3.28) this may be rewritten as

$$\downarrow e(T) = \sum_{\overline{T}_1=\overline{T}} \sum_{\overline{T}_2=\overline{T}} e(T_1)\frac{N(T)P(T)}{h(T)}e(T_2). \tag{2.3.37}$$

Since

$$e(T_1)\frac{N(T)P(T)}{h(T)}e(T_2) = e(\overline{T}_1)\frac{E(T_1)}{h(T_1)}e(\overline{T}_1)\frac{N(T)P(T)}{h(T)}e(\overline{T}_2)\frac{E(T_2)}{h(T_2)}e(\overline{T}_2)$$

from Theorem 2.1 we derive that T_1, T_2 and T must all have the same shape. But then the conditions $\overline{T}_1 = \overline{T} = \overline{T}_2$ force them all to be the same, converting (2.3.37) to

$$\downarrow e(T) = e(\overline{T})\frac{E(T)}{h(T)}e(\overline{T})\frac{N(T)P(T)}{h(T)}e(\overline{T})\frac{E(T)}{h(T)}e(\overline{T})$$

$$= e(\overline{T})\frac{P(T)N(T)}{h(T)}e(\overline{T})\frac{N(T)P(T)}{h(T)}e(\overline{T})\frac{P(T)N(T)}{h(T)}e(\overline{T})$$

(by (2.2.10) used twice) $= \dfrac{ab}{h(T)^3}e(\overline{T})P(T)N(T)\ P(T)N(T)e(\overline{T})$

(by (2.2.11)) $= \dfrac{ab}{h(T)^2}e(\overline{T})P(T)N(T)e(\overline{T}) = \dfrac{ab}{h(T)}e(T)$

$$\tag{2.3.38}$$

where

$$a = N(T)e(\overline{T})N(T)P(T)\Big|_\epsilon \quad \text{and} \quad b = e(\overline{T})P(T)N(T)P(T)\Big|_\epsilon.$$

To avoid computing these constants we simply observe that $\downarrow e(T)$ and $e(T)$ must have the same coefficient of the identity, and since this coefficient must be $1/h(T)$ by (2.3.27), the identity in (2.3.38) forces

$$ab = h(T)$$

and proves (2.3.36).

We are finally ready to prove seminormality.

Theorem 2.10 *For each partition λ we have constants*

$$d_1^\lambda, d_2^\lambda, \ldots, d_{n_\lambda}^\lambda \tag{2.3.39}$$

giving

$$\downarrow e_{ij}^\lambda = \frac{d_i^\lambda}{d_j^\lambda}\, e_{ji}^\lambda \qquad (for\ all\ 1 \le i, j \le n_\lambda). \tag{2.3.40}$$

Proof Since (2.3.40) may also be written as

$$\downarrow e_{ij}^\lambda = \frac{d_i^\lambda/d_1^\lambda}{d_j^\lambda/d_1^\lambda}\, e_{ji}^\lambda,$$

there is no loss in assuming that

$$d_1^\lambda = 1.$$

This given, we need only show that for some constants d_j we have

$$\downarrow e_{1j} = \frac{1}{d_j}\, e_{j1} \qquad (for\ 1 \le j \le n_\lambda), \tag{2.3.41}$$

where to lighten our notation we shall for a moment omit the superscript λ. In fact, (2.3.41) implies

$$\downarrow e_{i1} = d_i\, e_{1i} \qquad (for\ 1 \le i \le n_\lambda),$$

and then (2.3.23) gives

$$\downarrow e_{ij} = \downarrow(e_{i1}e_{1j}) = (\downarrow e_{1j})(\downarrow e_{i1}) = \frac{d_i}{d_j}\, e_{j1}\, e_{1i} = \frac{d_i}{d_j}\, e_{ji},$$

proving (2.3.40). So let us then prove (2.3.41). To this end we use the expansion formula (2.3.4) and get

$$\downarrow e_{1j} = \sum_{r=1}^{n_\lambda} \sum_{s=1}^{n_\lambda} \left((\downarrow e_{1j}) e_{sr} \big|_{\epsilon} \right) e_{rs} . \tag{2.3.42}$$

However, note that Theorem 2.9 gives

$$(\downarrow e_{1j}) e_{sr} \big|_{\epsilon} = e(\overline{T}_j) \frac{N(T_j) P(T_j)}{h_\lambda} \sigma_{j1} e(\overline{T}_1) e(\overline{T}_s) \frac{N(T_s) P(T_s)}{h_\lambda} \sigma_{sr} e(\overline{T}_r) \Big|_{\epsilon}$$

$$= \frac{N(T_j) P(T_j)}{h_\lambda} \sigma_{j1} e(\overline{T}_1) e(\overline{T}_s) \frac{N(T_s) P(T_s)}{h_\lambda} \sigma_{sr} e(\overline{T}_r) e(\overline{T}_j) \Big|_{\epsilon}$$

and the nonvanishing of the two products

$$e(\overline{T}_1) e(\overline{T}_s) \quad \text{and} \quad e(\overline{T}_r) e(\overline{T}_j)$$

forces $s = 1$ and $r = j$ reducing (2.3.42) to

$$\downarrow e_{1j} = \left((\downarrow e_{1j}) e_{1j} \big|_{\epsilon} \right) e_{j1}$$

and this proves (2.3.41) with

$$\frac{1}{d_j} = (\downarrow e_{1j}) e_{1j} \big|_{\epsilon} .$$

This completes the proof and this section. The actual nature of the constants in (2.3.39) will come to the surface in the next section.   □

2.4   The Murphy Elements

A remarkable set of group algebra elements was shown by Murphy [17] to play an elegant role in the study of representations of S_n. It develops that these elements considerably simplify manipulations with Young seminormal units. Their definition is quite simple. We set

$$m_k = (1, k) + (2, k) + (3, k) + \cdots + (k - 1, k) \qquad (\text{for } k = 2, 3, \ldots, n) . \tag{2.4.1}$$

These elements generate a commutative subalgebra of $\mathcal{A}(S_n)$. In fact we have the following basic relations.

Theorem 2.11 *Let s_h denote the transposition $(h, h + 1)$. Then*

$$
\begin{array}{lll}
(a) & m_h\, m_k = m_k\, m_h & \text{(for } 2 \leq h \leq k \leq n), \\
(b) & s_h\, m_k\, s_h\, m_k & \text{(for } h \neq k, k - 1), \\
(c) & s_k\, m_k\, s_k\, m_{k+1} - s_k, \\
(d) & s_{k-1}\, m_k\, s_{k-1}\, m_{k-1} + s_k.
\end{array}
\qquad (2.4.2)
$$

Proof Note first that in $\mathcal{A}(S_3)$ we have

$$(1, 2)\big((1, 2) + (1, 3) + (2, 3)\big) = \big((1, 2) + (1, 3) + (2, 3)\big)(1, 2)$$

and this immediately implies

$$m_2\, m_3 = m_3\, m_2.$$

This relation will clearly remain valid in $\mathcal{A}(S_n)$ for all $n \geq 3$. So we may proceed by induction and suppose that

$$m_2, \ m_3, \ \ldots, \ m_{n-1}$$

have been shown to commute in $\mathcal{A}(S_{n-1})$. Since they will necessarily commute also in $\mathcal{A}(S_n)$, and

$$C_2 = m_2 + m_3 + \cdots + m_n \qquad (2.4.3)$$

is a conjugacy class, we deduce that for all $2 \leq k \leq n - 1$ we have

$$
\begin{aligned}
m_2\, m_k + \cdots + m_{n-1}\, m_k + m_k\, m_n &= m_k\,(m_2 + m_3 + \cdots + m_n) \\
&= (m_2 + m_3 + \cdots + m_n)\, m_k \\
&= m_2\, m_k + \cdots + m_{n-1}\, m_k + m_n\, m_k.
\end{aligned}
$$

This yields

$$m_k\, m_n = m_n\, m_k \qquad \text{(for all } 2 \leq k \leq n - 1)$$

and proves (2.4.2) (a). Now note that (2.4.2) (b) is trivial when $h > k$. On the other hand, when $h < k$, conjugation of m_k by s_h only interchanges the terms (h, k) and $(h + 1, k)$ in m_k. Thus (2.4.2) (b) must hold true for all $h \neq k$ precisely as asserted. Finally, we see that we also have

$$s_k\big((1, k) + (2, k) + \cdots + (k - 1, k)\big)s_k$$
$$= (1, k + 1) + (2, k + 1) + \cdots + (k - 1, k + 1) - m_{k+1} - (k, k + 1).$$

This proves (2.4.2) (c). The proof is now complete since (2.4.2) (d) immediately follows from (2.4.2) (c) upon replacing k by $k - 1$.   □

What is truly remarkable is that the Murphy elements have the Young seminormal units e_{ij}^λ as their common eigenvectors. This is an immediate consequence of the following identity.

Theorem 2.12 *For every standard tableaux T of shape $\lambda \vdash n$ we have*

$$C_2 \, e(T) = \big(n(\lambda') - n(\lambda)\big) e(T), \tag{2.4.4}$$

where for a partition $\lambda = (\lambda_1, \lambda_2, \ldots, \lambda_k)$ we set

$$n(\lambda) = \sum_{i=1}^{k} (i - 1) \lambda_i \tag{2.4.5}$$

and λ' denotes the conjugate of λ.

Proof Since C_2 commutes with every element of $\mathcal{A}(S_n)$ we derive that

$$C_2 \, e(T) = e(\overline{T}) \frac{P(T) C_2 N(T)}{h_\lambda} e(\overline{T}), \tag{2.4.6}$$

and Von Neumann's lemma then yields

$$P(T) C_2 N(T) = C_2 N(T) P(T) \big|_\epsilon E(T).$$

Substituting this in (2.4.6) we get

$$C_2 \, e(T) = \Big(C_2 N(T) P(T) \big|_\epsilon \Big) \, e(\overline{T}) \frac{P(T) N(T)}{h_\lambda} e(\overline{T})$$

$$= \Big(C_2 N(T) P(T) \big|_\epsilon \Big) \, e(T).$$

It remains to prove

$$C_2 N(T) P(T) \big|_\epsilon = n(\lambda') - n(\lambda). \tag{2.4.7}$$

Now the definition in (2.2.1) gives

$$C_2 N(T) P(T) \big|_\epsilon = \sum_{1 \le i < j \le n} \sum_{\alpha \in R(T)} \sum_{\beta \in C(T)} \text{sign}(\beta) \chi \big(\beta\alpha = (i, j)\big). \tag{2.4.8}$$

However, it should be apparent that unless the indices i, j are in the same row or column of T it is impossible to interchange their positions in T by a row permutation followed by a column permutation without messing up the positions of the other entries in T. Thus the only way we can have the equality $\beta\alpha = (i, j)$ is $\beta = (i, j)$ or $\alpha = (i, j)$. This reduces the evaluation of the right hand side of (2.4.8) to counting transpositions in $C(T)$ and $R(T)$. Now if

$$\lambda = (\lambda_1, \lambda_2, \dots, \lambda_k) \quad \text{and} \quad \lambda' = (\lambda'_1, \lambda'_2, \dots, \lambda'_h)$$

then the number of transpositions in $C(T)$ and $R(T)$ are respectively given by

$$\sum_{i=1}^{h} \binom{\lambda'_i}{2} \quad \text{and} \quad \sum_{i=1}^{k} \binom{\lambda_i}{2}.$$

This reduces (2.4.8) to

$$C_2 N(T) P(T) \big|_{\epsilon} = \sum_{i=1}^{k} \binom{\lambda_i}{2} - \sum_{i=1}^{h} \binom{\lambda'_i}{2},$$

and it easily seen that this is only another way of writing the equality in (2.4.7). Our proof of (2.4.5) is thus complete. $\qquad\square$

Remark 2.2 We should mention that the identity in (2.4.7) implies a classical identity (see [16]) satisfied by the characters of S_n. To see this note that since conjugation by an element $\sigma \in S_n$ does not change the coefficient of the identity, formula (2.4.7) may also be rewritten as

$$\frac{h_\lambda}{n!} \sum_{\sigma \in S_n} C_2 \sigma \frac{N(T)P(T)}{h_\lambda} \sigma^{-1} \big|_{\epsilon} = n(\lambda') - n(\lambda).$$

But then Proposition 2.4 gives

$$\frac{h_\lambda}{n!} C_2 \chi^\lambda \big|_{\epsilon} = n(\lambda') - n(\lambda).$$

Since $n!/h_\lambda = n_\lambda$ we finally obtain

$$\frac{|C_2|}{n_\lambda} \chi^\lambda_{21^{n-2}} = n(\lambda') - n(\lambda), \tag{2.4.9}$$

where $\chi^\lambda_{21^{n-2}}$ gives the value of χ^λ at the conjugacy class C_2 and $|C_2|$ is the cardinality of C_2.

It is customary to call the difference $j - i$ the *content* of the lattice cell (i, j). For instance in figure (2.4.10) we display the Ferrers diagram of the partition $(5, 3, 3)$ with its cells filled by their contents.

-2	-1	0		
-1	0	1		
0	1	2	3	4

$$(2.4.10)$$

For a given tableau T with n cells and an integer $2 \leq k \leq n$ let us denote by $c_T(k)$ the content of the cell that contains k in T. This given it is easy to see that for any tableau T of shape λ we necessarily have

$$\sum_{k=2}^{n} c_T(k) = n(\lambda') - n(\lambda). \qquad (2.4.11)$$

This simple observation causes Theorem 2.12 to have the following truly remarkable corollary.

Theorem 2.13 *For every standard tableau T with n cells we have for $2 \leq k \leq n$*

$$(a)\ m_k\, e(T) = c_T(k)\, e(T),$$
$$(b)\ e(T)\, m_k = c_T(k)\, e(T). \qquad (2.4.12)$$

Proof Note that since by Theorem 2.9, $e(T)$ is self flipping and m_k is trivially so, (2.4.12) (b) can be obtained by flipping both sides of (2.4.12) (a). So we need only prove the latter. From the definition in (2.3.16) it follows that

$$e(\begin{smallmatrix}2\\1\end{smallmatrix}) = \epsilon - (1, 2) \quad \text{and} \quad e(\boxed{1\,2}) = \epsilon + (1, 2).$$

Thus we see that

$$m_2 e(\begin{smallmatrix}2\\1\end{smallmatrix}) = -e(\begin{smallmatrix}2\\1\end{smallmatrix}) \quad \text{and} \quad m_2\, e(\boxed{1\,2}) = e(\boxed{1\,2}).$$

This verifies (2.4.12) (a) for $n = 2$. So we may proceed by induction and suppose that (2.4.12) (a) has been verified up to $n - 1$. In particular if T is any standard tableau with n cells we will necessarily have

$$m_k\, e(\overline{T}) = c_{\overline{T}}(k)e(\overline{T}) = c_T(k)e(\overline{T}) \qquad \text{(for } 2 \leq k \leq n-1). \qquad (2.4.13)$$

Thus

$$m_k\, e(T) = m_k\, e(\overline{T})\, \frac{P(T)N(T)}{h(T)}\, e(\overline{T})$$

$$\text{(by (2.4.13))} = c_T(k)\, e(\overline{T})\, \frac{P(T)N(T)}{h(T)}\, e(\overline{T}) = c_T(k)\, e(T). \qquad (2.4.14)$$

This proves (2.4.12) for $2 \leq k \leq n-1$. But for $k = n$ we may use (2.4.4) with $n(\lambda') - n(\lambda)$ given by (2.4.11) and get

$$(m_2 + m_3 + \cdots + m_n)\, e(T) = (c_T(2) + c_T(3) + \cdots + c_T(n))\, e(T).$$

Subtracting the identities in (2.4.14) for $2 \leq k \leq n-1$ yields

$$m_n\, e(T) = c_T(n)\, e(T),$$

completing the induction and the proof.   □

A beautiful consequence of this result is a completely explicit formula for Young's seminormal units. To present this development we need a few preliminary observations. To begin note that if $P(x_2, x_3, \ldots, x_n)$ is any polynomial in its arguments and T is any standard tableau, then from Theorem 2.13 it follows that

$$P(m_2, m_3, \ldots, m_n)\, e(T) = P(c_T(2), c_T(3), \ldots, c_T(n))\, e(T).$$

Now we may construct $P(x_2, x_3, \ldots, x_n)$ in such as manner that

$$P(c_T(2), c_T(3), \ldots, c_T(n)) = 1$$

while at the same time the operator $P(m_2, m_3, \ldots, m_n)$ kills all the other seminormal idempotents. To give a precise and efficient construction of such a polynomial we need notation.

Let us recall that the *addable cells* of a partition μ of $n-1$ are the cells we may add to the Ferrers diagram of μ to obtain of the Ferrers diagram of a partition of n. The collection of contents of the addable cells of μ will be simply denoted $\mathcal{A}C_\mu$. For instance it is easily seen from (2.4.10) that

$$\mathcal{A}C_{533} = \{-3, 2, 5\}.$$

This given we have

Theorem 2.14 *For any standard tableau T we recursively define a polynomial $P_T(x_2, x_3, \ldots, x_n)$ by setting*

$$P_T(x_2, x_3, \ldots, x_n) \tag{2.4.15}$$

$$= P_{\overline{T}}(x_2, x_3, \ldots, x_{n-1}) \prod_{\substack{c \in \mathcal{A}C_{\lambda(\overline{T})} \\ c \neq c_T(n)}} \frac{x_n - c}{c_T(n) - c} \quad \left(\textit{with } P_{[1]} = 1\right).$$

Then

$$P_T(m_2, m_3, \ldots, m_n)\, e(S) = \begin{cases} 0 & \text{if } S \text{ is standard and } S \neq T, \\ e(T) & \text{if } S = T. \end{cases} \tag{2.4.16}$$

Proof For $T = [1]$ there is nothing to prove. We can thus proceed by induction and assume (2.4.16) to be valid for all for tableaux with $n - 1$ cells. Note that the induction hypothesis immediately implies that the expression

$$P_{\overline{T}}(m_2, m_3, \ldots, m_{n-1})\, e(\overline{S})$$

fails to vanish only if and only if $\overline{S} = \overline{T}$ and in that case we have

$$P_{\overline{T}}(m_2, m_3, \ldots, m_{n-1})\, e(\overline{T}) = 1.$$

This also proves the first case of (2.4.16) when $\overline{S} \neq \overline{T}$. So we are left with the case $\overline{S} = \overline{T}$. Now from from (2.4.15) it follows that for $\overline{S} = \overline{T}$

$$P_T(m_2, m_3, \ldots, m_n)\, e(S) = \left(\prod_{\substack{c \in \mathcal{A}C_{\lambda(\overline{T})} \\ c \neq c_T(n)}} \frac{x_k - c}{c_T(n) - c} \right) e(S)$$

$$\big(\text{by } (2.4.12)\,(a)\big) = \left(\prod_{\substack{c \in \mathcal{A}C_{\lambda(\overline{T})} \\ c \neq c_T(n)}} \frac{c_S(n) - c}{c_T(n) - c} \right) e(S).$$

This may also be written as

$$P_T(m_2, m_3, \ldots, m_n)\, e(S) = \left(\prod_{\substack{c \in \mathcal{A}C_\mu \\ c \neq c_T(n)}} \frac{c_S(n) - c}{c_T(n) - c} \right) e(S) \tag{2.4.17}$$

where for convenience we have set

$$\mu = \lambda(\overline{S}) = \lambda(\overline{T}).$$

Note further that the equality $\overline{S} = \overline{T}$ forces $c_S(n) \in \mathcal{A}C_\mu$. So for the right hand side of (2.4.17) not to vanish we must have $c_S(n) = c_T(n)$. But this holds true if and only if $S = T$. In this case we have

$$\prod_{\substack{c \in \mathcal{A}C_{\lambda(\overline{T})} \\ c \neq c_T(n)}} \frac{c_S(n) - c}{c_T(n) - c} = \prod_{\substack{c \in \mathcal{A}C_{\lambda(\overline{T})} \\ c \neq c_T(n)}} \frac{c_T(n) - c}{c_T(n) - c} = 1$$

and (2.4.17) reduces to

$$P_T(m_2, m_3, \ldots, m_n)\, e(T) = e(T),$$

completing the proof of the Theorem. □

We can now prove the following remarkable fact.

Theorem 2.15 *For any standard tableau T we have*

$$e(T) = P_T(m_2, m_3, \ldots, m_n). \tag{2.4.18}$$

Proof Assume that

$$T = T_r^\mu. \tag{2.4.19}$$

Then

$$
\begin{aligned}
P_T(m_2, m_3, \ldots, m_n) e_{ji}^\lambda \big|_\epsilon
&= P_T(m_2, m_3, \ldots, m_n) e_{jj}^\lambda e_{ji}^\lambda e_{ii}^\lambda \big|_\epsilon \\
&= P_T(m_2, m_3, \ldots, m_n) e(T_j^\lambda) e_{ji}^\lambda\, e(T_i^\lambda) \big|_\epsilon \\
&= e(T_j^\lambda) e_{ji}^\lambda\, e(T_i^\lambda)\, P_T(m_2, m_3, \ldots, m_n) \big|_\epsilon
\end{aligned}
\tag{2.4.20}
$$

Now, by (2.4.19) and Theorem 2.14, we have

$$P_T(m_2, m_3, \ldots, m_n) e(T_j^\lambda) \neq 0 \quad \Longrightarrow \quad \lambda = \mu \ \text{ and } \ j = r. \tag{2.4.21}$$

Similarly, in view of (2.4.12) (b) we also derive that

$$e(T_i^\lambda)\, P_T(m_2, m_3, \ldots, m_n) \neq 0 \quad \Longrightarrow \quad \lambda = \mu \ \text{ and } \ i = r. \tag{2.4.22}$$

Using (2.4.20), (2.4.21), and (2.4.22) the expansion in (2.3.4) with $f = P_T(m_2, m_3, \ldots, m_n)$ reduces to

$$
\begin{aligned}
P_T(m_2, m_3, \ldots, m_n) &= h_\mu\, P_T(m_2, m_3, \ldots, m_n) e(T_r^\mu) \big|_\epsilon e(T_r^\mu) \\
\text{(by (2.4.19))} &= h_\mu\, P_T(m_2, m_3, \ldots, m_n) e(T) \big|_\epsilon e(T) \\
\text{(by Theorem 2.14)} &= h_\mu\, e(T) \big|_\epsilon e(T) \\
\text{(by (2.3.27))} &= e(T)
\end{aligned}
$$

completing the proof of the Theorem. □

It might be good at this point to exhibit a few instances of the identity in (2.4.18). For the tableaux with three cells Theorem 2.14 gives

$$e(\boxed{\begin{array}{ccc}1&2&3\end{array}}) = (m_2 + 1)(m_3 + 1)/6, \quad e(\boxed{\begin{array}{c}3\\1\ 2\end{array}}) = (m_2 + 1)(m_3 - 2)/6$$

$$e(\boxed{\begin{array}{c}3\\2\\1\end{array}}) = -(m_2 - 1)(m_3 - 1)/6, \quad e(\boxed{\begin{array}{c}2\\1\ 3\end{array}}) = -(m_2 - 1)(m_3 + 2)/6.$$

We should also note that we have

$$e(\boxed{\begin{array}{c}4\\2\ 5\\1\ 3\end{array}}) = -(m_2 - 1)(m_3 + 2)(m_4 - 2)(m_4 + 2)(m_5 - 2)/96.$$

These expressions are easily derived if we take the view that the successive linear factors must be selected to kick each additional label into the position that it occupies in the target tableau.

2.5 The Seminormal Matrices

In this section we shall work out explicit formulas for the seminormal matrices corresponding to simple reflections for any given partition μ. Since we shall keep μ fixed throughout our derivation, to simplify the displays we will sometimes omit the superscript μ. So we assume that

$$T_1, T_2, \ldots, T_{n_\mu} \tag{2.5.1}$$

are the standard tableaux of shape μ in the Young's last letter order. We pick a simple transposition $s_k = (k, k+1)$ (for $1 \leq k \leq n-1$) and proceed to construct all the entries of the seminormal matrix $A^\mu(s_k)$. Suppose first that for a pair $1 \leq r < s \leq n_\mu$ we have

$$s_k T_r = T_s. \tag{2.5.2}$$

Note that since T_r and T_s only differ in the positions of k and $k+1$, the last letter of disagreement between T_r and T_s is necessarily $k + 1$. Moreover, since $r < s \Rightarrow T_r <_{LL} T_s$ it follows that $k + 1$ is higher in T_r than in T_s. Clearly k and $k + 1$ cannot be in the same row or column of T_r for then (2.5.2) would force T_s not to be standard. Furthermore two successive integers can never be in the same diagonal of any standard tableau. In conclusion we see that these two tableaux can only be as indicated in figure (2.5.3).

$$(2.5.3)$$

This given our first step is to compute the image by s_k of the seminormal idempotent $e_{rr} = e(T_r)$. To this end we use the expansion formula in (2.3.4) and start by writing

$$s_k e(T_r) = \sum_\lambda h_\lambda \sum_{i,j=1}^{n_\lambda} \left(s_k e(T_r) e_{ji}^\lambda \Big|_\epsilon \right) e_{ij}^\lambda . \qquad (2.5.4)$$

However, since by our conventions $e(T_r) = e_{rr}^\mu$ a use of Theorem 2.6 quickly reduces (2.5.4) to

$$
\begin{aligned}
s_k e(T_r) &= h_\mu \sum_{i=1}^{n_\mu} \left(s_k e(T_r) e_{ri}^\mu \Big|_\epsilon \right) e_{ir}^\mu \\
&= h_\mu \sum_{i=1}^{n_\mu} \left(s_k e_{ri}^\mu \Big|_\epsilon \right) e_{ir}^\mu \qquad (2.5.5) \\
&= h_\mu \sum_{i=1}^{n_\mu} e_{ri}^\mu (s_k) e_{ir}^\mu .
\end{aligned}
$$

Now it develops that this expansion can be reduced dramatically further by use of Murphy elements. To this end set

$$\Pi = \prod_{\substack{h=2 \\ h \neq k,k+1}}^{n} \prod_{\substack{c=c_1 \\ c \neq c_{T_r}(h)}}^{c_2} \frac{(m_h - c)}{(c_{T_r}(h) - c)} \qquad (2.5.6)$$

where c_1 and c_2 respectively denote the minimum and the maximum of the contents of the cells of the diagram of μ. Note that since T_r and T_s only differ in the positions of k and $k+1$ we will have

$$c_{T_r}(h) = c_{T_s}(h) \qquad \text{(for all } h \neq k, k+1). \qquad (2.5.7)$$

It follows from (2.5.6) and (2.4.12) (a) that

$$\Pi e(T_i) = \begin{cases} 0 & \text{if } i \neq r, s, \\ e(T_r) & \text{if } i = r, \\ e(T_s) & \text{if } i = s. \end{cases} \qquad (2.5.8)$$

In fact, the last two cases of (2.5.8) are immediate consequences of (2.5.6) and (2.5.7). On the other hand we see from (2.5.6) that

$$\Pi\, e(T_i) \neq 0 \implies c_{T_i}(h) = c_{T_r}(h) \quad \text{for all} \quad h \neq k, k+1\,.$$

Since standard tableaux increase along diagonals these equalities force T_i to be identical with T_r except for the positions of k and $k + 1$. In other words the nonvanishing of $\Pi e(T_i)$ forces $T_i = T_r$ or $T_i = T_s$. This proves the first case of (2.5.8).

Note next that, since there is no occurrence of m_k and m_{k+1} in the right hand side of (2.5.6), it follows from Theorem 2.11 that

$$\Pi\, s_k = s_k\, \Pi\,. \tag{2.5.9}$$

We get

$$s_k\, e(T_r) = s_k\, \Pi\, e(T_r)$$

$$(\text{by } (2.5.9)) = \Pi\, s_k e(T_r)$$

$$(\text{by } (2.5.5)) = h_\mu \sum_{i=1}^{n_\mu} e_{ri}^\mu(s_k)\, \Pi\, e_{ir}^\mu$$

$$(\text{by } (2.3.23)) = h_\mu \sum_{i=1}^{n_\mu} e_{ri}^\mu(s_k)\, \Pi\, e_{ii}^\mu\, e_{ir}^\mu$$

$$(\text{by } (2.5.9)) = h_\mu\, e_{rr}^\mu(s_k)\, e_{rr}^\mu + h_\mu\, e_{rs}^\mu(s_k)\, e_{sr}^\mu\,.$$

This may be rewritten as

$$s_k\, e_{rr}^\mu = a\, e_{rr}^\mu + b\, e_{sr}^\mu\,, \tag{2.5.10}$$

where a and b are constants we shall soon determine.

To begin we multiply both sides of (2.5.10) by the Murphy element m_k and get (using (2.4.2) (c))

$$(s_k\, m_{k+1} - 1)\, e_{rr}^\mu = a\, m_k\, e_{rr}^\mu + b\, m_k\, e_{sr}^\mu\,,$$

and (2.4.12) (a) gives

$$c_{T_r}(k+1)\, s_k\, e_{rr}^\mu - e_{rr}^\mu = a\, c_{T_r}(k)\, e_{rr}^\mu + b\, c_{T_r}(k+1)\, e_{sr}^\mu\,.$$

Subtracting from this (2.5.10) multiplied by $c_{T_r}(k+1)$ yields

$$-e_{rr}^\mu = a\, (c_{T_r}(k) - c_{T_r}(k+1))\, e_{rr}^\mu\,.$$

Thus

$$a = \frac{1}{c_{T_r}(k+1) - c_{T_r}(k)}. \qquad (2.5.11)$$

Our next step is to multiply (2.5.10) by s_k. This gives

$$e_{rr}^\mu = a\, s_k\, e_{rr}^\mu + b\, s_k\, e_{sr}^\mu.$$
$$\text{(by (2.5.10))} = a(a\, e_{rr}^\mu + b\, e_{sr}^\mu) + b\, s_k\, e_{sr}^\mu \qquad (2.5.12)$$
$$= a^2\, e_{rr}^\mu + ab\, e_{sr}^\mu + b\, s_k\, e_{sr}^\mu.$$

Note that $c_{T_r}(k+1) - c_{T_r}(k) = 1$ forces k and $k+1$ to be in the same row of T_r and $c_{T_r}(k+1) - c_{T_r}(k) = -1$ forces k and $k+1$ to be in the same column. Since these two alternatives have been excluded we cannot have $a^2 = 1$. But then (2.5.12) shows that we cannot have $b = 0$. This allows us to extract $s_k\, e_{sr}^\mu$ out of (2.5.12) and obtain

$$s_k\, e_{sr}^\mu = \frac{(1-a^2)}{b}\, e_{rr}^\mu - a\, e_{sr}^\mu. \qquad (2.5.13)$$

To determine b we need a further consequence of Theorem 2.3 which is quite interesting in its own right. This may be stated as follows.

Proposition 2.7 *For two standard tableaux $T_1\ T_2$ of the same shape we have*

$$(a)\ \ E(T_2)\, e(T_1) \neq 0 \ \ \Longrightarrow \ \ T_2 <_{LL} T_1,$$
$$(b)\ \ e(T_1)\, E(T_2) \neq 0 \ \ \Longrightarrow \ \ T_1 <_{LL} T_2.$$

Proof Let $k+1$ be the last letter of disagreement between T_1 and T_2. In view of the recursive definition of the seminormal unit $e(T_2)$ given in (2.3.16) we may write $e(T_1)$ in the form

$$e(T_1) = e(T_1(k-1))P(T_1(k))\, R_1 \qquad (2.5.14)$$

where R_1 is a residual factor of no concern. In the same vein, using (2.2.15) (b) we may write

$$E(T_2) = P(T_2)\, n_k(T_2)N(T_2(k)). \qquad (2.5.15)$$

Using (2.5.14) and (2.5.15) we see that

$$E(T_2)\, e(T_1) \neq 0 \ \ \Longrightarrow \ \ N(T_2(k))e(T_1(k-1))P(T_1(k)) \neq 0.$$

Thus from Proposition 2.1 it follows that

$$E(T_2)\,e(T_1) \neq 0 \implies \lambda\big(T_2(k)\big) \;\geq\; \lambda\big(T_1(k)\big) \quad \text{(in dominance)}$$

and part (a) of the proposition necessarily follows from the definition of Young's last letter order. The proof of part (b) is entirely analogous. □

As a corollary of Proposition 2.7 we can now immediately derive the following surprising result.

Proposition 2.8 *The constant b appearing in (2.5.10) and (2.5.13) is plainly and simply equal to* 1.

Proof Note that since $e_{sr}^{\mu} = e_{ss}^{\mu} e_{sr}^{\mu} e_{rr}^{\mu}$ we may write

$$h_\mu\, e_{sr}^{\mu} = e(\overline{T}_s)\frac{E(T_s)}{h_\mu}e(\overline{T}_s)\Big(e(\overline{T}_s)E(T_s)\sigma_{sr}^{\mu}e(\overline{T}_r)\Big)e(\overline{T}_r)\frac{E(T_r)}{h_\mu}e(\overline{T}_r)$$

$$= e(\overline{T}_s)\frac{E(T_s)}{h_\mu}\Big(e(\overline{T}_s)E(T_s)\sigma_{sr}^{\mu}e(\overline{T}_r)\Big)\frac{E(T_r)}{h_\mu}e(\overline{T}_r)$$

$$= e(T_s)E(T_s)\sigma_{sr}^{\mu}e(T_r)$$

and since by assumption we have $s_k\,T_r = T_s$ we can set $\sigma_{sr}^{\mu} = s_k$ in this last identity and obtain

$$h_\mu\, e_{sr}^{\mu} = e(T_s)E(T_s)s_k\,e(T_r)$$

$$(\text{by (2.5.10)}) = e(T_s)E(T_s)\big(a\,e_{rr}^{\mu} + b\,e_{sr}^{\mu}\big)$$

$$= a\,e(T_s)E(T_s)e(T_r) + b\,e(T_s)E(T_s)e_{sr}^{\mu}$$

$$= a\,e(T_s)E(T_s)e(T_r) + b\,e(T_s)E(T_s)\Big(e(T_s)\frac{E(T_s)}{h_\mu}s_k\,e(T_r)\Big)$$

$$(\text{Using (2.3.17) (c)}) = a\,e(T_s)E(T_s)e(T_r) + b\,e(T_s)E(T_s)s_k\,e(T_r)$$

$$= a\,e(T_s)E(T_s)e(T_r) + b\,h_\mu\,e_{sr}^{\mu}\,.$$

$$(2.5.16)$$

Now note that we cannot have

$$E(T_s)e(T_r) \neq 0$$

for otherwise our Proposition 2.7 would give $s < r$ which contradicts our original assumptions. Thus (2.5.16) necessarily reduces to

$$h_\mu\, e_{sr}^{\mu} = b\,h_\mu\,e_{sr}^{\mu}$$

which forces

$$b = 1$$

and completes our proof.   □

To continue our construction of the seminormal matrix corresponding to the simple transposition s_k we need to compute the image by s_k of the idempotent $e(T_r)$ when k and $k+1$ are in the same row or column. Our point of departure, also in these cases is formula (2.5.5). Moreover, since now the tableau $s_k\,T_r$ is no longer standard, from (2.5.6) we derive that

$$\Pi\,e(T_i) = \begin{cases} 0 & \text{if } i \neq r, \\ e(T_r) & \text{if } i = r. \end{cases} \qquad (2.5.17)$$

This given, multiplying both sides of (2.5.5) by Π we obtain

$$s_k e(T_r) = s_k \Pi\, e(T_r) = \Pi\, s_k e(T_r) = h_\mu \sum_{i=1}^{n_\mu} e_{ri}^\mu(s_k)\, \Pi\, e_{ir}^\mu$$

$$= h_\mu \sum_{i=1}^{n_\mu} e_{ri}^\mu(s_k)\, \Pi\, e_{ii}^\mu\, e_{ir}^\mu$$

$$(\text{by } (2.5.17)) = h_\mu\, e_{rr}^\mu(s_k) e_{rr}^\mu$$

$$= a\, e(T_r)$$

with

$$a = h_\mu\, e_{rr}^\mu(s_k). \qquad (2.5.18)$$

To determine a we multiply both sides of (2.5.18) by the Murphy element m_k and use (2.4.2) (c) to get

$$a\, c_{T_r}(k)\, e_{rr}^\mu = m_k\, s_k e_{rr}^\mu$$

$$= s_k m_{k+1}\, e(T_r) - e(T_r)$$

$$= c_{T_r}(k+1)\, s_k\, e(T_r) - e(T_r)$$

$$(\text{by } (2.5.18)) = c_{T_r}(k+1)\, a\, e(T_r) - e(T_r),$$

and this gives

$$a = \frac{1}{c_{T_r}(k+1) - c_{T_r}(k)} = \begin{cases} 1 & \text{if } k \text{ and } k+1 \text{ are in the same row of } T_r, \\ -1 & \text{if } k \text{ and } k+1 \text{ are in the same column.} \end{cases} \qquad (2.5.19)$$

To state our final result in a compact form we need one further observation concerning the case when k and $k + 1$ are not in the same row or column of T_r. Then if $k + 1$ is in cell (i_1, j_1) and k is in cell (i_2, j_2) we may write

$$c_{T_r}(k) - c_{T_r}(k + 1) = (j_2 - i_2) - (j_1 - i_1)$$
$$= j_2 - j_1 + i_1 - i_2 .$$

When $k + 1$ is in a higher cell than k in T_r, as it was assumed at the beginning of this section, then the quantity

$$\pi_k(T_r) = c_{T_r}(k) - c_{T_r}(k + 1) = -1/a \qquad (2.5.20)$$

gives the *taxi-cab distance* between k and $k + 1$. That is the length of any path that joins $k + 1$ to k by *EAST* and *SOUTH* steps. In figure (2.5.21) we have drawn such a path by a solid fat line.

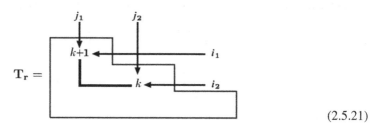

$$(2.5.21)$$

This given, the identities proved in this section yield the following fundamental result.

Theorem 2.16 *Let*

$$T_1^\mu, T_2^\mu, \ldots, T_{n_\mu}^\mu$$

be the standard tableaux of shape $\mu \vdash n$ in Young's last letter order. Then for any pair $1 \le r, i \le n_\mu$ and $1 \le k < n$ we have

1. $s_k\, e_{ru}^\mu = e_{ru}^\mu$ *if k and $k + 1$ are in the same row of T_r^μ,* $\qquad (2.5.22)$

2. $s_k\, e_{ru}^\mu = -e_{ru}^\mu$ *if k and $k + 1$ are in the same column of T_r^μ,*

3. $s_k\, e_{ru}^\mu = -\dfrac{1}{\pi_k(T_r^\mu)}\, e_{ru}^\mu + e_{su}^\mu$ *if $T_s^\mu = s_k T_r^\mu$ and $k + 1$ is higher*

   than k in T_r^μ,

4. $s_k\, e_{ru}^\mu = \dfrac{1}{\pi_k(T_r^\mu)}\, e_{ru}^\mu + \left(1 - \dfrac{1}{\pi_k^2(T_r^\mu)}\right) e_{su}^\mu$ *if $T_s^\mu = s_k T_r^\mu$ and k is higher*

   than $k + 1$ in T_r^μ.

Proof Given (2.5.19), the first two cases are simply a restatement of (2.5.17) right multiplied by e_{ru}^{μ}. Note next that if we combine (2.5.10), (2.5.11), (2.5.20) and Proposition 2.7 we obtain

$$s_k\, e_{rr}^{\mu} = -\frac{1}{\pi_k(T_r^{\mu})}\, e_{rr}^{\mu} + e_{sr}^{\mu}\,. \tag{2.5.23}$$

This gives the third case after right multiplication by e_{ru}^{μ}. Similarly, (2.5.13) gives

$$s_k\, e_{sr}^{\mu} = \left(1 - \frac{1}{\pi_k^2(T_r^{\mu})}\right) e_{rr}^{\mu} + \frac{1}{\pi_k(T_r^{\mu})}\, e_{sr}^{\mu}\,. \tag{2.5.24}$$

Since $\pi_k(T_r^{\mu}) = \pi_k(T_s^{\mu})$, the fourth case of (2.5.22) is obtained by an interchange of r and s followed by right multiplication by e_{su}^{μ}. This completes the proof. $\square$

From (2.5.22) we obtain a rather simple recipe for constructing the seminormal representation matrix $A^{\mu}(s_k)$. More precisely we have the following result.

Theorem 2.17 *If $A^{\mu}(s_k)$ is the matrix yielding the action of the simple transposition $s_k = (k, k+1)$ on the basis $\langle e_{1,u}^{\mu}, e_{2,u}^{\mu}, \ldots, e_{n_{\mu},u}^{\mu}\rangle$, that is*

$$s_k\langle e_{1,u}^{\mu}, e_{2,u}^{\mu}, \ldots, e_{n_{\mu},u}^{\mu}\rangle = \langle e_{1,u}^{\mu}, e_{2,u}^{\mu}, \ldots, e_{n_{\mu},u}^{\mu}\rangle A^{\mu}(s_k)\,, \tag{2.5.25}$$

then $A^{\mu}(s_k) = \|a_{ij}^{\mu}(s_k)\|_{i,j=1}^{n_{\mu}}$ with

$$a_{rr}(s_k) = \begin{cases} \dfrac{-1}{\pi_k(T_r^{\mu})} & \text{if } k+1 \text{ is higher than } k \text{ in } T_r^{\mu}, \\[4mm] \dfrac{1}{\pi_k(T_r^{\mu})} & \text{if } k \text{ is higher than } k+1 \text{ in } T_r^{\mu}, \end{cases} \tag{2.5.26}$$

where $\pi_k(T_r^{\mu})$ is the taxi-cab distance between k and $k+1$ in T_r^{μ}. Moreover for $i \neq j$ we have

$$a_{ij}(s_k) = \begin{cases} 0 & \text{if } s_k T_i^{\mu} \neq T_j^{\mu}, \\[2mm] \dfrac{1}{1 - \pi_k^2(T_i^{\mu})} & \text{if } s_k T_i^{\mu} = T_j^{\mu} \text{ and } i < j, \\[3mm] 1 & \text{if } s_k T_i^{\mu} = T_j^{\mu} \text{ and } i > j\,. \end{cases} \tag{2.5.27}$$

Proof The equation in (2.5.24) simply says that for each $1 \leq j \leq n_{\mu}$ we have

$$s_k e_{ju}^{\mu} = \sum_{i=1}^{n_{\mu}} e_{iu}^{\mu}\, a_{ij}^{\mu}(s_k)\,.$$

Thus the identities in (2.3.26) and (2.3.27) are obtained by equating coefficients of the units e_{iu}^μ on both sides of (2.5.22). □

We are now in a position to obtain explicit expressions for the factors d_i^μ occurring in (2.3.40). Our starting point is the following auxiliary identity.

Proposition 2.9 Let $T_s^\mu = s_k\, T_r^\mu$ with $r < s$ then

$$d_s^\mu/d_r^\mu = \left(1 - 1/\pi_k^2(T_r)\right). \tag{2.5.28}$$

Proof Equating coefficients of the identity in (2.5.24) and using (2.3.27) gives

$$s_k\, e_{sr}^\mu\big|_\epsilon = \left(1 - \frac{1}{\pi_k^2(T_r)}\right)\frac{1}{h_\mu}. \tag{2.5.29}$$

On the other hand we have

$$s_k\, e_{sr}^\mu\big|_\epsilon = \downarrow(s_k\, e_{sr}^\mu)\big|_\epsilon$$

$$= (\downarrow e_{sr}^\mu)\, s_k\big|_\epsilon$$

$$\text{(by (2.3.40))} = \frac{d_s^\mu}{d_r^\mu}\, e_{rs}^\mu\, s_k\big|_\epsilon$$

$$= \frac{d_s^\mu}{d_r^\mu}\, s_k\, e_{rs}^\mu\big|_\epsilon$$

$$\text{(by (2.5.23) right multiplied by } e_{rs}^\mu) = \frac{d_s^\mu}{d_r^\mu}\left(-1/\pi_k(T_r^\mu)\, e_{rs}^\mu + e_{ss}^\mu\right)\Big|_\epsilon$$

$$= \frac{d_s^\mu}{d_r^\mu}\frac{1}{h_\mu}\,, \tag{2.5.30}$$

and (2.5.28) follows by combining (2.5.29) and (2.5.30). □

To complete the construction of these factors we need one further result.

Proposition 2.10 For any λ and any $1 \le s \le n_\lambda$ we can join T_1^λ to T_s^λ by a chain of tableaux

$$T_1^\lambda = T_{i_1}^\lambda \to T_{i_2}^\lambda \to \cdots \to T_{i_{m-1}}^\lambda \to T_{i_m}^\lambda = T_s^\lambda \tag{2.5.31}$$

with the following two basic properties:

(a) $T_{i_{r-1}}^\lambda <_{LL} T_{i_r}^\lambda$ for all $1 \le r \le m-1$,

(b) $T_{i_r}^\lambda = s_{k_r}\, T_{i_{r-1}}^\lambda$ with $s_{k_r} = (k_r, k_r + 1)$. $\tag{2.5.32}$

Proof The assertion is trivial for $\lambda \vdash 2$. So, by induction, let us assume we have proved the result for any $\mu \vdash n - 1$. Let then $\lambda \vdash n$ and $1 \leq s \leq n_\lambda$. To construct the chain in (2.5.31) we need some notation. To begin let c_s be the lattice cell that contains n in T_s^λ and let c_1 be the lattice cell that contains n in T_1^λ. We should note that c_1 is necessarily be the highest corner of the diagram of λ. Now set $\lambda(\overline{T}_s^\lambda) = \mu$. Using the induction hypothesis we construct a chain for $\overline{T}_s^\lambda$:

$$T_1^\mu = T_{j_1}^\mu \to T_{j_2}^\mu \to \cdots \to T_{j_{m-1}}^\mu \to T_{j_m}^\mu = \overline{T}_s^\lambda . \tag{2.5.33}$$

Let $T_{i_r}^\lambda$, for $1 \leq r \leq m$, be the tableau obtained by adding n to $T_{j_r}^\mu$ in the cell c_s. If it happens that $c_1 = c_s$ then the tableaux $T_{i_r}^\lambda$ will be satisfy (2.5.31) and (2.5.32) and we are done. If c_s is a lower corner of the diagram of λ then the chain

$$T_{i_1}^\lambda \to T_{i_2}^\lambda \to \cdots \to T_{i_{m-1}}^\lambda \to T_{i_m}^\lambda = \overline{T}_s^\lambda \tag{2.5.34}$$

will satisfy (2.5.32) (a) and (b). Now we only need to construct a chain that joins T_1^λ to $T_{i_1}^\lambda$. The crucial observation is that when $c_s \neq c_1$ then c_1 is the highest corner of the diagrams of both μ and λ. Thus in T_1^μ the label $n - 1$ is necessarily in c_1. In particular $T_{i_1}^\lambda$ has n in c_s and $n - 1$ in c_1. This given, the tableau $s_{n-1} T_{i_1}^\lambda$ will have n in c_1 and $n - 1$ in c_s. Thus

$$s_{n-1} T_{i_1}^\lambda <_{LL} T_{i_1}^\lambda .$$

This reduces us to the previous case since now n is the highest corner of λ. Thus the construction of the desired chain can now be carried out by prepending the chain in (2.5.34) with the chain that joins T_1^λ to $s_{n-1} T_{i_1}^\lambda$. This completes the induction and the proof.   □

Combining the last two propositions we obtain the following.

Theorem 2.18 *If*

$$T_1^\lambda = T_{i_1}^\lambda \to T_{i_2}^\lambda \to \cdots \to T_{i_{m-1}}^\lambda \to T_{i_m}^\lambda = T_s^\lambda$$

is any chain satisfying (2.5.32) (a) and (b) then

$$d_s^\lambda = \prod_{r=1}^{m-1} \left(1 - \pi_{k_r}^2 (T_{i_r}^\mu)\right) . \tag{2.5.35}$$

Proof Clearly, (2.5.35) follows by successive applications of (2.5.28).   □

2.6 Murphy Elements and Conjugacy Classes

We should point out that what we now refer to as the Murphy elements actually first appeared in a 1971 paper of A. Jucys, only to be rediscovered by Murphy in 1980. Murphy's fundamental contribution is discovering their connection with Young's seminormal units. However, in a remarkable 1972 paper Jucys goes on to show that that every class function of S_n may be expressed as a symmetric polynomial in $m_2, m_3, \ldots, m_n$. Jucys' development was basically existential. The purpose of this section is to reestablish Jucys' results in a constructive manner and obtain explicit expressions for the characters of S_n as well as the conjugacy classes. We should mention that such a constructive approach was adopted in a joint paper by Diaconis and Greene [6] where a number of explicit formulas were derived in special cases. We also solve here a number of problems posed therein.

To begin we shall derive two separate proofs of the following basic fact.

Theorem 2.19 *If $Q(y_2, y_3, \ldots, y_n)$ is a symmetric polynomial in its arguments then the group algebra element*

$$Q(m_2, m_3, \ldots, m_n) \tag{2.6.1}$$

is a class function of S_n.

Proof The result follows if we prove that

$$\sigma \, Q(m_2, m_3, \ldots, m_n) \sigma^{-1} = Q(m_2, m_3, \ldots, m_n) \quad \text{(for all } \sigma \in S_n) . \tag{2.6.2}$$

Since the simple transpositions $s_1, s_2, \ldots, s_{n-1}$ generate S_n we need only check the identities

$$s_k \, Q(m_2, m_3, \ldots, m_n) \, s_k = Q(m_2, m_3, \ldots, m_n) \quad \text{(for } k = 1, 2, \ldots, n-1) . \tag{2.6.3}$$

Moreover, since every symmetric polynomial in $y_2, y_3, \ldots, y_n$ is a polynomial in the elementary symmetric functions

$$e_1(y_2, y_3, \ldots, y_n), e_2(y_2, y_3, \ldots, y_n), \ldots, e_{n-1}(y_2, y_3, \ldots, y_n) , \tag{2.6.4}$$

and products of class functions are class function, we need only verify (2.6.2) when Q is one of the elementaries in (2.6.4).

Now we can do this all at once by showing (2.6.3) for

$$Q(y_2, y_3, \ldots, y_n) = \sum_{r=1}^{n-1} t^r \, e_r(y_2, y_3, \ldots, y_n) = \prod_{h=2}^{n} (1 + t \, y_h) .$$

To this end note that the relations in (2.4.2) (b) give us

$$s_k \left(\prod_{h=2}^{n} (1 + t \, m_h) \right) s_k = \left(\prod_{\substack{h=2 \\ h \neq k, k-1}}^{n} (1 + t \, m_h) \right) s_k (1 + t m_k)(1 + t m_{k+1}) s_k . \tag{2.6.5}$$

But from (2.4.2) (c) and (d) we then obtain

$$s_k \, (1 + t \, m_k) \, (1 + t \, m_{k+1}) s_k$$
$$= s_k \, (1 + t \, m_k) \, s_k \, s_k \, (1 + t \, m_{k+1}) s_k \tag{2.6.6}$$
$$= (1 + t \, m_{k+1} - t \, s_k) \, (1 + t \, m_k + t \, s_k)$$
$$= (1 + t \, m_{k+1} - t \, s_k + t \, m_k + t \, s_k + t^2 (m_{k+1} - s_k)(m_k + s_k)$$
$$= (1 + t \, m_{k+1} + t \, m_k + t^2 (m_{k+1} m_k + m_{k+1} s_k - s_k \, m_k - 1) .$$

Now note that right multiplication of (2.4.2) (c) by s_k also gives

$$m_{k+1} s_k - s_k \, m_k - 1 = 0 .$$

Substituting this in (2.6.6) reduces it to

$$s_k \, (1 + t \, m_k) \, (1 + t \, m_{k+1}) s_k = (1 + t \, m_{k+1} + t \, m_k + t^2 (m_{k+1} m_k)$$
$$= (1 + t \, m_k) \, (1 + t \, m_{k+1}) .$$

Using this relation in (2.6.5) finally yields the desired identity

$$s_k \left(\prod_{h=2}^{n} (1 + t \, m_h) \right) s_k = \left(\prod_{h=2}^{n} (1 + t \, m_h) \right)$$

completing our first proof.   ●

We have seen that the seminormal units $e(T)$ may be expressed as polynomials in the Murphy elements. We need to do the reverse here and express the Murphy elements in terms of the seminormal units. More precisely,

Theorem 2.20 *For $k \leq n$ we have*

$$m_k = \sum_{T \in ST(n)} c_T(k) \, e(T) \tag{2.6.7}$$

where the symbol $T \in ST(n)$ is to indicate that the sum is to be carried out over all standard tableaux with n cells.

Proof Formula (2.3.4) with $f = m_k$ gives

$$m_k = \sum_{\lambda \vdash n} h_\lambda \sum_{i,j=1}^{n_\lambda} m_k \, e_{ji}^\lambda \big|_\epsilon \, e_{ij}^\lambda . \tag{2.6.8}$$

However from the relations in (2.3.23) we derive that

$$m_k \, e_{ji}^\lambda = m_k \, e_{jj}^\lambda e_{ji}^\lambda$$

$$(\text{by } (2.4.12)\,(a)) = c_{T_j^\lambda}(k) \, e_{jj}^\lambda e_{ji}^\lambda = c_{T_j^\lambda}(k) \, e_{ji}^\lambda ,$$

and (2.3.27) gives

$$m_k \, e_{ji}^\lambda \big|_\epsilon = c_{T_j^\lambda}(k) \, e_{ji}^\lambda \big|_\epsilon = \begin{cases} 0 & \text{if } i \neq j \\ c_{T_j^\lambda}(k) \, / h_\lambda & \text{if } i = j . \end{cases}$$

Using this in (2.6.8) reduces it to

$$m_k = \sum_{\lambda \vdash n} \sum_{j=1}^{n_\lambda} c_{T_j^\lambda}(k) \, e_{jj}^\lambda .$$

This completes our proof of Theorem 2.20 since this identity is simply another way of writing (2.6.7). $\qquad \Box$

We are now ready to give an alternate proof of Theorem 2.19.

Second Proof From (2.6.7) and the orthogonality relations in (2.3.20) we derive that

$$m_{k_1} m_{k_2} \cdots m_{k_l} = \sum_{T \in ST(n)} c_T(k_1) \, c_T(k_2) \cdots c_T(k_l) \, e(T) .$$

It thus follows that for any polynomial $Q(y_2, y_2, \ldots, y_n)$ we must also have

$$Q(m_2, m_3, \cdots, m_n) = \sum_{T \in ST(n)} Q\big(c_T(2), c_T(3), \ldots, c_T(n)\big) \, e(T) . \tag{2.6.9}$$

Now if $Q(y_2, y_2, \ldots, y_n)$ is a symmetric function in its arguments, the order in which the quantities $c_T(2), c_T(3), \ldots, c_T(n)$ are substituted in $Q(y_2, y_2, \ldots, y_n)$ is immaterial and by grouping terms where $Q\big(c_T(2), c_T(3), \ldots, c_T(n)\big)$ takes the same value, formula (2.6.9) may be rewritten in the form

$$Q(m_2 m_3 \cdots m_n) = \sum_{\lambda \vdash n} Q\big(c_\lambda(2), c_\lambda(3), \ldots, c_\lambda(n)\big) \sum_{T \in ST(\lambda)} e(T) . \tag{2.6.10}$$

where the quantities $c_\lambda(2), c_\lambda(3), \ldots, c_\lambda(n)$ represent the contents of the cells of the diagram of λ in some preferred order and the symbol $T \in ST(\lambda)$ represents that the sum is to be carried out over all standard tableaux of shape λ. Now (2.6.10) would complete the proof once we realize that the summand

$$U^\lambda = \sum_{T \in ST(\lambda)} e(T) = \sum_{i=1}^{n_\lambda} e_{ii}^\lambda \qquad (2.6.11)$$

is none other than the character χ^λ. However, all we need here is to show that the U^λ are class functions, and for that we need only verify that they commute with all the seminormal units e_{rs}^μ. However this is an immediate consequence of the relations in (2.3.23).   •

To show the converse of Theorem 2.19 we need only show that the irreducible characters χ^λ may be expressed as symmetric polynomials in the Murphy elements. To see where this leads us, suppose that for each partition λ we have a symmetric polynomial

$$Q_\lambda(x_2, x_3, \ldots, x_n) \qquad (2.6.12)$$

such that

$$\chi^\lambda = Q_\lambda(m_2, m_3, \ldots, m_n) . \qquad (2.6.13)$$

Now it follows immediately from Proposition 2.5 and (2.3.34) that

$$\chi^\lambda = h_\lambda \sum_{i=1}^{n_\lambda} e_{ii}^\lambda . \qquad (2.6.14)$$

Using the identities in (2.3.23 we then immediately derive that

$$\chi^\lambda \chi^\mu = \begin{cases} 0 & \text{if } \mu \neq \lambda, \\ h_\lambda \chi^\lambda & \text{if } \mu = \lambda. \end{cases} \qquad (2.6.15)$$

These are the well known orthogonality relations satisfied by the irreducible characters. Combining (2.6.13) with (2.6.15) we derive that we must have

$$Q_\lambda(m_2, m_3, \ldots, m_n) \chi^\mu = \begin{cases} 0 & \text{if } \mu \neq \lambda, \\ h_\lambda \chi^\lambda & \text{if } \mu = \lambda. \end{cases} \qquad (2.6.16)$$

However, from (2.4.12) (a) and the symmetry of Q_λ it follows (as in the second proof of Theorem 2.19) that

$$Q_\lambda(m_2, m_3, \ldots, m_n) \, e_{ii}^\mu = Q_\lambda(c_\mu(2), c_\mu(3), \ldots, c_\mu(n)) \, e_{ii}^\mu$$

and thus we also have

$$Q_\lambda(m_2, m_3, \ldots, m_n) \, \chi^\mu = Q_\lambda(c_\mu(2), c_\mu(3), \ldots, c_\mu(n)) \, \chi^\mu . \tag{2.6.17}$$

Comparing with (2.6.16) we derive that the polynomial Q_λ must satisfy the identities

$$Q_\lambda(c_\mu(2), c_\mu(3), \ldots, c_\mu(n)) = \begin{cases} 0 & \text{if } \mu \neq \lambda, \\ h_\lambda & \text{if } \mu = \lambda. \end{cases} \tag{2.6.18}$$

Experimenting with special cases shows that these equations do not uniquely determine Q_λ. Nevertheless, all we need here is a systematic way of constructing one particular solution for each λ. Now it develops that we may in fact, produce Q_λ in terms of a symmetric polynomial which first appears in an exercise of Macdonald (see [16, Ex. 5, 6 and 7, p. 117]).

Let us recall that the symbol $(x)_a$ represents the *lower factorial polynomial* that is

$$(x)_a = x(x - 1)(x - 2) \cdots (x - a + 1) .$$

Our starting point is the following remarkable result.

Proposition 2.11 *For a given integer vector* $a = (a_1 > a_2 > \cdots > a_n \geq 0)$ *put* $|a| = a_1 + a_2 + \cdots + a_n$ *and set*

$$\Xi_a(y) = \Xi_a(y_1, y_2, \ldots, y_n) = \frac{\det \left\| (y_i)_{a_j} \right\|_{i,j=1}^n}{\prod_{1 \leq i < j \leq n}(y_i - y_j)} . \tag{2.6.19}$$

Then for all $b = (b_1 > b_2 > \cdots > b_n \geq 0)$ *we have*

$$\Xi_a[b] = \begin{cases} \dfrac{a_1! a_2! \cdots a_n!}{\prod_{1 \leq i < j \leq n}(a_i - a_j)} & \text{if } b = a, \\ 0 & \text{if } |b| \leq |a| \text{ and } b \neq a. \end{cases} \tag{2.6.20}$$

Proof By the definition of determinant we have

$$\det \left\| (y_i)_{a_j} \right\|_{i,j=1}^n = \sum_{\sigma \in S_n} \text{sign}(\sigma) \, (y_1)_{a_{\sigma_1}} (y_2)_{a_{\sigma_2}} \cdots (y_n)_{a_{\sigma_n}} . \tag{2.6.21}$$

Now note that for an integer x we have

$$(x)_a = \frac{x!}{(x-a)!}.$$

Thus with the replacements $y_i \to b_i$ we may rewrite (2.6.21) in the form

$$\det \left\| (b_i)_{a_j} \right\|_{i,j=1}^n = b_1! b_2! \cdots b_n! \tag{2.6.22}$$

$$\times \sum_{\sigma \in S_n} \mathrm{sign}(\sigma) \frac{1}{(b_1 - a_{\sigma_1})!(b_2 - a_{\sigma_2})! \cdots (b_n - a_{\sigma_n})!}.$$

For the summand corresponding to σ to survive we must have

$$b_1 \geq a_{\sigma_1}, \ b_2 \geq a_{\sigma_2}, \ \ldots, \ b_n \geq a_{\sigma_n},$$

and this gives

$$b_1 + b_2 + \cdots + b_n \geq a_1 + a_2 + \cdots + a_n.$$

Thus when $|b| \leq |a|$ all terms in (2.6.21) do vanish unless $b = a$. But in this case only the identity term survives, and (2.6.22) reduces to

$$\det \left\| (a_i)_{a_j} \right\|_{i,j=1}^n = a_1! a_2! \cdots a_n!.$$

This implies (2.6.20) precisely as stated.   □

The identities in (2.6.20) strongly suggest that we should be able to use the polynomial $\Xi_a(y)$ in the construction of Q_λ. This is precisely what we will do. However to carry this out we need some notation and a few auxiliary facts.

For a pair of integers $m, k \geq 0$ set

$$\Sigma_k(m) = \sum_{s=1}^m i^k, \tag{2.6.23}$$

with the understanding that

$$\Sigma_0(m) = m. \tag{2.6.24}$$

The following result is well known.

Proposition 2.12 *The sequence of polynomials $R_k(x)$ with generating function*

$$\sum_{k \geq 0} \frac{u^k}{k!} R_k(x) = e^u \frac{e^{ux} - 1}{e^u - 1} \tag{2.6.25}$$

yields all the integer sums in (2.6.23). More precisely we have

$$\Sigma_k(m) = R_k(m) \qquad \text{for all } k, m \geq 0 .$$ (2.6.26)

Proof We have

$$\sum_{k \geq 0} \frac{u^k}{k!} \Sigma_k(m) = \sum_{k \geq 0} \frac{u^k}{k!} \sum_{i=1}^{m} i^k$$

$$= \sum_{i=1}^{m} \sum_{k \geq 0} \frac{(iu)^k}{k!} = \sum_{i=1}^{m} e^{iu}$$

$$= e^u + e^{2u} + \cdots + e^{mu}$$

$$= e^u \frac{e^{mu} - 1}{e^u - 1} .$$

This proves (2.6.26). $\qquad\qquad\qquad\qquad\qquad\qquad\qquad\qquad\qquad$ $\square$

It will be good to see here a few of these polynomials. We should mention that they were computed with MAPLE directly from (2.6.25):

$$R_1(x) = \frac{1}{2}x(x+1)$$

$$R_2(x) = \frac{1}{6}x(x+1)(2x+1)$$

$$R_3(x) = \frac{1}{4}x^2(x+1)^2$$

$$R_4(x) = \frac{1}{30}x(2x+1)(x+1)(3x^2+3x-1)$$

$$R_5(x) = \frac{1}{12}x^2(2x^2++2x-1)(x+1)^2$$

$$R_6(x) = \frac{1}{42}x(2x+1)(x+1)(3x^4+6x^3-3x+1) .$$

If $\lambda = (\lambda_1 \geq \lambda_2 \geq \cdots \geq \lambda_m \geq 0)$ is a partition we set $l(\lambda) = m$ and call it the *length of* λ. If $l(\lambda) \leq n$ we set

$$\lambda(n) = \big(\lambda_1(n), \lambda_2(n), \ldots, \lambda_n(n)\big)$$

with

$$\lambda_i(n) = \begin{cases} \lambda_i + n - 1 & \text{if } 1 \leq i \leq l(\lambda) \\ n - i & \text{if } l(\lambda) < i \leq n . \end{cases}$$

It will also be convenient here and after to denote by $C(\lambda)$ the sequence of contents of the diagram of a partition λ, in some order.

Recalling that we denote by $p_k(x_1, x_2, \ldots, x_n)$ the *power sum* symmetric function

$$p_k(x) = p_k(x_1, x_2, \ldots, x_n) = x_1^k + x_2^k + \cdots + x_n^k$$

we have the following basic fact

Theorem 2.21 *The symmetric polynomial*

$$\pi_{n,k}(x) = R_k(n-1) + \sum_{s=1}^{k} \binom{k}{s} \left(n^s - (n-1)^s\right) p_{k-s}(x) \tag{2.6.27}$$

yields the identities

$$p_k[\lambda(n)] = \pi_{n,k}[C(\lambda)] \qquad (\text{for all } \lambda \vdash n) . \tag{2.6.28}$$

Proof If $l(\lambda) < n$ set

$$\lambda_i = 0 \qquad (\text{for all } i > l(\lambda)) .$$

From the definition of contents we then get that

$$
\begin{aligned}
p_k[C(\lambda)] &= \sum_{i=1}^{n} \sum_{j=1}^{\lambda_i} (j - i)^k \\
&= \sum_{i=1}^{n} \sum_{j=1}^{\lambda_i} \sum_{a=0}^{k} \binom{k}{a} (-i)^{k-a} j^a \tag{2.6.29} \\
&= \sum_{i=1}^{n} \sum_{a=0}^{k} \binom{k}{a} (-i)^{k-a} R_a(\lambda_i) .
\end{aligned}
$$

Note that this formula remains valid for $k = 0$ provided we adopt the convention of setting

$$p_0[C(\lambda)] = n \qquad (\text{for all } \lambda \vdash n) . \tag{2.6.30}$$

In fact, setting $k = 0$, the last line of (2.6.29) reduces to

$$\sum_{i=1}^{n} R_0(\lambda_i)$$

and (2.6.24) gives

$$\sum_{i=1}^{n} R_0(\lambda_i) = \sum_{i=1}^{n} \lambda_i = n.$$

This given, we derive that

$$\sum_{k\geq 0} \frac{p_k[C(\lambda)]}{k!} u^k = \sum_{i=1}^{n} \sum_{a\geq 0} \frac{R_a(\lambda_i)}{a!} u^a \sum_{k\geq a} \frac{(-i)^{k-a}}{(k-a)!} u^{k-a}$$

$$(\text{using } (2.6.26)) = \sum_{i=1}^{n} \frac{e^u}{e^u - 1} \left(e^{\lambda_i u} - 1\right) e^{-i u}$$

$$= \frac{1}{1 - e^{-u}} \sum_{i=1}^{n} \left(e^{(\lambda_i - i) u} - e^{-i u}\right)$$

$$= \frac{e^{-n u}}{1 - e^{-u}} \sum_{i=1}^{n} \left(e^{(\lambda_i + n - i) u} - e^{(n-i)u}\right) \qquad (2.6.31)$$

$$= \frac{e^{-n u}}{1 - e^{-u}} \sum_{i=1}^{n} \sum_{k\geq 1} \left(\lambda_i^k(n) - (n-i)^k\right) \frac{u^k}{k!}$$

$$= \frac{e^{-n u}}{1 - e^{-u}} \sum_{k\geq 1} \left(p_k[\lambda(n)] - R_k(n-1)\right) \frac{u^k}{k!}.$$

Now this may be inverted to

$$\sum_{k\geq 1} p_k[\lambda(n)] \frac{u^k}{k!} = \sum_{k\geq 1} R_k(n-1) \frac{u^k}{k!} + \left(e^{n u} - e^{(n-1)u}\right) \sum_{k\geq 0} p_k[C(\lambda)] \frac{u^k}{k!}.$$

Equating coefficients of u^k in this equality gives

$$p_k[\lambda(n)] = R_k(n-1) + \sum_{s=1}^{k} \binom{k}{s} \left(n^s - (n-1)^s\right) p_{k-s}[C(\lambda)].$$

and this proves (2.6.28) with $\pi_{n,k}$ given by (2.6.27), precisely as asserted.   □

This result has the following immediate corollary.

Theorem 2.22 *Let P be a symmetric polynomial in $(x_1, x_2, \ldots, x_n)$ and let*

$$P = H(p_1, p_2, \ldots, p_n) = \sum_{\rho} c_\rho p_\rho$$

be its expansion in terms of the power symmetric function basis. Then the symmetric polynomial

$$\pi_n P = H(p_1, p_2, \cdots, p_n)\Big|_{p_k \to \pi_{n,k}}$$

yields the identities

$$(\pi_n P)[C(\lambda)] = P[\lambda(n)] \qquad \text{(for all } \lambda \vdash n) .$$

Now we need only one more fact to obtain our polynomial Q_λ, namely the following classical formula.

Lemma 2.2 *The number of standard tableaux of shape $\lambda \vdash n$ is given by*

$$f_\lambda = \frac{n!}{\lambda_1(n)!\lambda_2(n)!\cdots\lambda_n(n)!} \prod_{1 \leq i < j \leq n} (\lambda_i(n) - \lambda_j(n)) . \qquad (2.6.32)$$

In particular we get that

$$h_\lambda = \frac{n!}{f_\lambda} = \frac{\lambda_1(n)!\lambda_2(n)!\cdots\lambda_n(n)!}{\prod_{1 \leq i < j \leq n} (\lambda_i(n) - \lambda_j(n))} . \qquad (2.6.33)$$

Proof It is well known (see [16, p. 114]) that for any $\rho \vdash n$ we have the Schur function expansion

$$p_\rho = \sum_{\lambda \vdash n} \chi_\rho^\lambda s_\lambda .$$

Since we also have

$$s_\lambda(x_1, x_2, \ldots, x_n) = \frac{\det \left\| x_i^{\lambda_j + n - j} \right\|_{i,j=1}^n}{\det \left\| x_i^{n-j} \right\|_{i,j=1}^n}$$

we see that

$$\chi_\rho^\lambda = p_\rho \det \left\| x_i^{n-j} \right\|_{i,j=1}^n \Big|_{x_1^{\lambda_1(n)} x_2^{\lambda_2(n)} \cdots x_n^{\lambda_n(n)}} . \qquad (2.6.34)$$

In the particular case that $\rho = 1^n$ this gives

$$
\begin{aligned}
f_\lambda &= \sum_{\sigma \in S_n} \text{sign}(\sigma) \sum_{\alpha_1 + \cdots + \alpha_n = n} \frac{n!}{\alpha_1! \cdots \alpha_n!} x_1^{\alpha_1 + n - \sigma_1} \cdots x_n^{\alpha_n + n - \sigma_n} \Big|_{x_1^{\lambda_1(n)} \cdots x_n^{\lambda_n(n)}} . \\
&= n! \sum_{\sigma \in S_n} \text{sign}(\sigma) \frac{1}{(\lambda_1(n) + \sigma_1 - n)!(\lambda_2(n) + \sigma_2 - n)! \cdots (\lambda_n(n) + \sigma_n - n)!} \\
&= \frac{n!}{\lambda_1(n)! \lambda_2(n)! \cdots \lambda_n(n)!} \sum_{\sigma \in S_n} \text{sign}(\sigma) \left(\lambda_1(n)\right)_{n-\sigma_1} \left(\lambda_2(n)\right)_{n-\sigma_2} \cdots \left(\lambda_n(n)\right)_{n-\sigma_n} \\
&= \frac{n!}{\lambda_1(n)! \lambda_2(n)! \cdots \lambda_n(n)!} \det \left\| \left(\lambda_i(n)\right)_{n-j} \right\|_{i,j=1}^n .
\end{aligned}
$$

$$(2.6.35)$$

This yields the desired formula (2.6.32) since the determinant in (2.6.35) can be reduced by simple column manipulations to the Vandermonde determinant evaluated at $\lambda_1(n), \lambda_2(n), \ldots, \lambda_n(n)$. □

We can now finally state the main result of this section.

Theorem 2.23 *For* $\lambda \vdash n$ *set*

$$Q_\lambda[y_1, y_2, \ldots, y_n] = \pi_n \, \Xi_{\lambda(n)} [y_1, y_2, \ldots, y_n] . \tag{2.6.36}$$

Then for $\mu \vdash n$ *we have*

$$Q_\lambda[C(\mu)] = \begin{cases} h_\lambda & \text{if } \mu = \lambda \\ 0 & \text{if } \mu \neq \lambda . \end{cases} \tag{2.6.37}$$

In particular we must also have

$$\chi^\lambda = Q_\lambda[0, m_2, \ldots, m_n] . \tag{2.6.38}$$

Proof Since the numerator in (2.6.19) is clearly an alternating polynomial in $y_1, y_2, \ldots, y_n$, the Vandermonde determinant factors out and the ratio evaluates to a symmetric polynomial. We can thus apply Theorem 2.22 and obtain

$$Q_\lambda[C(\mu)] = \Xi_{\lambda(n)}[\mu(n)] .$$

Since

$$\sum_{i=1}^n \lambda_i(n) = \binom{n}{2} + n = \sum_{i=1}^n \mu_i(n)$$

we can apply Proposition 2.11 and derive that

$$
Q_\lambda[C(\mu)] = \begin{cases} \dfrac{\lambda_1(n)!\lambda_2(n)!\cdots\lambda_n(n)!}{\prod_{1\le i<j\le n}\left(\lambda_i(n)-\lambda_j(n)\right)} & \text{if } \mu \ne \lambda\,, \\ 0 & \text{if } \mu = \lambda \end{cases}
$$

and (2.6.37) immediately follows from the identity in (2.6.33).

Now note that combining (2.6.37) with (2.4.12) we get for any standard tableau T of shape μ

$$
Q_\lambda[0, m_2, \ldots, m_n]\, e(T) = Q_\lambda(C(\mu))\, e(T) = \begin{cases} h_\lambda\, e(T) & \text{if } \mu = \lambda \\ 0 & \text{if } \mu \ne \lambda\,. \end{cases}
$$

$$(2.6.39)$$

Since

$$
\chi^\mu = \sum_{\lambda(T)=\mu} e(T)
$$

from (2.6.39) we get

$$
Q_\lambda[0, m_2, \ldots, m_n]\, \chi^\mu = \begin{cases} h_\lambda\, \chi^\lambda & \text{if } \mu = \lambda\,, \\ 0 & \text{if } \mu \ne \lambda\,. \end{cases}
$$

Subtracting from this result the well known identities

$$
\chi^\lambda\, \chi^\mu = \begin{cases} h_\lambda\, \chi^\lambda & \text{if } \mu = \lambda\,, \\ 0 & \text{if } \mu \ne \lambda \end{cases}
$$

yields

$$
\left(Q_\lambda[0, m_2, \ldots, m_n] - \chi^\lambda \right) \chi^\mu = 0\,, \qquad \text{(for all } \mu \vdash n)\,. \qquad (2.6.40)
$$

Since Theorem 2.19 guarantees that the group algebra element

$$
Q_\lambda[0, m_2, \ldots, m_n] - \chi^\lambda
$$

is a class function, the identities in (2.6.40) are sufficient to guarantee that this difference must identically vanish. This completes our proof.   □

It develops that these polynomials are quite remarkable in their relative simplicity. To begin with, computer experimentation reveals that the polynomial $\Xi_{\lambda(n)}[y_1, y_2, \ldots, y_n]$ may be quite monstrous even for small partitions. Of course that should be expected. When expressed in terms of the power basis it may still take a few lines of print even for $\lambda \vdash 4$. However, surprisingly, the replacement $p_k \to \pi_{n,k}$ produces dramatic simplifications so that the resulting polynomials end up containing only a few terms. For instance for the partitions of 4 we obtain

$$Q_{1111} = 1 + 4p_1 - p_2/2 + p_1^2/2 - p_3$$

$$Q_{211} = -3 - 6p_1 + 3p_2/2 - p_1^2/2 + p_3$$

$$Q_{22} = 20 + + p_1 - 4p_2$$

$$Q_{31} = -3 + 6p_1 + 3p_2/2 - p_1^2/2 - p_3$$

$$Q_4 = 1 - 4p_1 - p_2/2 + p_1^2/2 + p_3 \ .$$

Even for partitions of 6 these polynomials contain only a few terms. For instance we get

$$Q_{321} = 161 - 22p_2 - 8p_1^2 - p_2^2 + 6p_4 \ .$$

Theorem 2.23 has two corollaries that are worth stating at this point.

Theorem 2.24 *The polynomial*

$$H_n[y_1, y_2, \ldots, y_n] = \sum_{\lambda \vdash n} Q_\lambda[y_1, y_2, \ldots, y_n]$$

satisfies the identities

$$H_n[C(\mu)] = h_\mu \qquad (\text{for all } \mu \vdash n). \qquad (2.6.41)$$

Proof This is an immediate consequence of (2.6.37). □

Again these polynomials are surprisingly simple. For instance up to $n = 6$ we obtain

$$H_3(y) = 1 + p_2$$

$$H_4(y) = 16 - 2p_2 + p_1^2$$

$$H_5(y) = p_4 - 6p_2 - p_1^2 + 46$$

$$H_6(y) = -9p_4 + 2p_2^2 + 14p_2 + 12p_1^2 + 11$$

Theorem 2.25 *For $\rho \vdash n$ set*

$$Z_\rho[y_1, y_2, \ldots, y_n] = \sum_{\lambda \vdash n} Q_\lambda[y_1, y_2, \ldots, y_n] \chi_\rho^\lambda / z_\rho \qquad (2.6.42)$$

where for $\rho = 1^{\alpha_1} 2^{\alpha_2} \cdots n^{\alpha_n}$ as customary we set

$$z_\rho = 1^{\alpha_1} 2^{\alpha_2} \cdots n^{\alpha_n} \alpha_1! \alpha 2! \cdots \alpha_n! \,. \tag{2.6.43}$$

Then

$$\mathcal{Z}_\rho[C(\mu)] = h_\mu \, \chi_\rho^\mu / z_\rho \tag{2.6.44}$$

and the conjugacy class C_ρ (as an element of the group algebra of S_n) is given by the formula

$$C_\rho = \mathcal{Z}_\rho[0, m_2, \ldots, m_n] \,. \tag{2.6.45}$$

Proof The identity in (2.6.44) follows immediately from (2.6.37). Next, recall that the class function C_ρ, in terms of the characters, has the expansion

$$C_\rho = \sum_{\lambda \vdash n} \chi^\lambda \, \chi_\rho^\lambda / z_\rho \,.$$

Thus (2.6.45) is an immediate consequence of (2.6.38).   □

We give below a few samples we obtained from formula (2.6.42)

$$\mathcal{Z}_{[2,2]}(y) = \frac{1}{2} p_1^2 - \frac{3}{2} p_2$$

$$\mathcal{Z}_{[3,2]}(y) = 12 p_1 + p_1 p_2 - 4 p_3$$

$$\mathcal{Z}_{[3,3]}(y) = -30 + 6 p_2 + 3 p_1^2 + 6 p_2 + \frac{1}{2} p_2^2 - \frac{5}{2} p_4 \,.$$

In a recent paper [13] A. Goupil et al. endeavour to express the central parameter

$$\omega_\rho^\lambda = \chi_\rho^\lambda h_\lambda / z_\rho \tag{2.6.46}$$

as a polynomial depending on ρ evaluated at $C(\lambda)$. The results they obtain are interesting in the present context since their polynomials yield alternate versions of the polynomials $\mathcal{Z}_\rho$. More precisely one of their results may be stated as follows

Theorem 2.26 (Goupil et al. [13]) *For each partition of n of the form*

$$\gamma, 1^{n-r} \qquad \text{with } \gamma \vdash r$$

we can construct a symmetric polynomial

$$G_\gamma = \sum_{|\rho| \leq r} c_\rho^\gamma(n) \, p_\rho \qquad \text{with } c_\rho(n) \in \mathbb{Q}[n] \tag{2.6.47}$$

satisfying

$$G_\gamma[C(\lambda)] = \sum_{|\rho| \leq r} c_\rho^\gamma(n)\, p_\rho[C(\lambda)] = \omega_{\gamma,1^{n-r}}^\lambda \qquad \textit{(for all } \lambda \vdash n) .$$

$$(2.6.48)$$

The polynomials G_γ constructed by Goupil et al. are also relatively simple when expressed in terms of power sums. The proof of Theorem 2.26 by Goupil et al. is algorithmic, but the resulting algorithm is of considerable complexity. Our aim here is to obtain an alternate algorithm.

It develops that the crucial idea in carrying this out stems from a calculation initiated by Macdonald in the same previously quoted exercise [16, Ex. 7, p. 117]. In fact, the aim of Macdonald in that exercise is to obtain a formula for $\omega_{k,1^{n-k}}^\lambda$. Extending Macdonald's idea to its most general form naturally leads us to the following remarkable result.

Theorem 2.27 *Let γ be a partition of m all of whose parts are > 1 and set*

$$\Psi_{\gamma,n}(y) = \frac{1}{z_\gamma} \sum_{\alpha \vdash r} \chi_\gamma^\alpha\, \Xi_{\alpha(n)}(y) . \qquad (2.6.49)$$

Then the polynomial

$$\Phi_{\gamma,n}(y) = \pi_n\, \Psi_{\gamma,n}(y) \qquad (2.6.50)$$

yields the identities

$$\Phi_{\gamma,n}\big[C(\lambda)\big] = w_{\gamma,1^{n-r}}^\lambda \qquad \textit{(for all } \lambda \vdash n) . \qquad (2.6.51)$$

Proof From the well known formula

$$\chi_\rho^\lambda = p_\rho\, \det \big\| x_i^{n-j} \big\|_{i,j=1}^n \Big|_{x_1^{\lambda_1+n-1} x_2^{\lambda_2+n-2} \cdots x_n^{\lambda_n+n-n}}$$

using our notation, we get for $\rho = \gamma,\, 1^{n-r}$

$$\chi_{\gamma,1^{n-r}}^\lambda = p_1^{n-r}\, p_\gamma\, \det \big\| x_i^{n-j} \big\|_{i,j=1}^n \Big|_{x_1^{\lambda_1(n)} x_2^{\lambda_2(n)} \cdots x_n^{\lambda_n(n)}} .$$

Expanding p_γ in terms of Schur functions we get

$$\chi_{\gamma,1^{n-r}}^\lambda = \sum_{\alpha \vdash r} \chi_\gamma^\alpha\, p_1^{n-r} s_\alpha(x_1,\ldots,x_n) \Delta_n(x_1,\ldots,x_n) \Big|_{x_1^{\lambda_1(n)} x_2^{\lambda_2(n)} \cdots x_n^{\lambda_n(n)}}$$

$$= \sum_{\alpha \vdash r} \chi_\gamma^\alpha\, p_1^{n-r} \det \big\| x_i^{\alpha_j(n)} \big\|_{i,j=1}^n \Big|_{x_1^{\lambda_1(n)} x_2^{\lambda_2(n)} \cdots x_n^{\lambda_n(n)}}$$

$$= \sum_{\alpha \vdash r} \chi_\gamma^\alpha\, p_1^{n-r} \sum_{\sigma \in S_n} \text{sign}(\sigma)\, x_1^{\alpha_{\sigma_1}(n)} x_2^{\alpha_{\sigma_2}(n)} \cdots x_n^{\alpha_{\sigma_n}(n)} \Big|_{x_1^{\lambda_1(n)} x_2^{\lambda_2(n)} \cdots x_n^{\lambda_n(n)}} .$$

Following Macdonald we use the multinomial expansion of p_1^{n-r} and obtain

$$\chi_{\gamma,1^{n-r}}^{\lambda} = \sum_{\alpha \vdash r} \chi_{\gamma}^{\alpha} \sum_{\sigma \in S_n} \text{sign}(\sigma) \sum_{p_1+\cdots+p_n=n-r} \frac{(n-r)!}{p_1! \cdots p_n!} x_1^{\alpha_{\sigma_1}(n)+p_1} \cdots x_n^{\alpha_{\sigma_n}(n)+p_n} \Big|_{x_1^{\lambda_1(n)} \cdots x_n^{\lambda_n(n)}}$$

$$= (n-r)! \sum_{\alpha \vdash r} \chi_{\gamma}^{\alpha} \sum_{\sigma \in S_n} \text{sign}(\sigma) \frac{1}{(\lambda_1(n) - \alpha_{\sigma_1}(n))! \cdots (\lambda_n(n) - \alpha_{\sigma_n}(n))!} \, .$$

Since all the parts of γ are > 1 we have that $z_{\gamma,1^{n-r}} = z_\gamma (n-r)!$ thus (2.6.46) with $\rho = \gamma,\, 1^{n-r}$ and (2.6.33) give

$$\omega_{\gamma,1^{n-r}}^{\lambda} = \chi_{\gamma,1^{n-r}}^{\lambda} \frac{h_\lambda}{z_\gamma (n-r)!} = \frac{\chi_{\gamma,1^{n-r}}^{\lambda}}{z_\gamma (n-r)!} \frac{\lambda_1(n)! \lambda_2(n)! \cdots \lambda_n(n)!}{\prod_{1 \le i < j \le n} (\lambda_i(n)! - \lambda_j(n)!)} \, .$$

Using this in our previous relation we obtain

$$w_{\gamma,1^{n-r}}^{\lambda} = \frac{1}{z_\gamma \prod_{1 \le i < j \le n} (\lambda_i(n)! - \lambda_j(n)!)}$$

$$\times \sum_{\alpha \vdash r} \chi_{\gamma}^{\alpha} \sum_{\sigma \in S_n} \text{sign}(\sigma) \frac{\lambda_1(n)! \lambda_2(n)! \cdots \lambda_n(n)!}{(\lambda_1(n) - \alpha_{\sigma_1}(n))! \cdots (\lambda_n(n) - \alpha_{\sigma_n}(n))!}$$

$$= \frac{1}{z_\gamma} \sum_{\alpha \vdash r} \chi_{\gamma}^{\alpha} \frac{\sum_{\sigma \in S_n} \text{sign}(\sigma)\, \lambda_1(n))_{\alpha_{\sigma_1}} (\lambda_2(n))_{\alpha_{\sigma_2}} \cdots (\lambda_n(n))_{\alpha_{\sigma_n}}}{\prod_{1 \le i < j \le n} (\lambda_i(n)! - \lambda_j(n)!)}$$

$$= \frac{1}{z_\gamma} \sum_{\alpha \vdash r} \chi_{\gamma}^{\alpha}\, \Xi_{\alpha(n)}[\lambda(n)] \, .$$

$$(2.6.52)$$

Now, from (2.6.50) and Theorem 2.22 we get that

$$\Phi_{\gamma,n}[C(\lambda)] = \Psi_{\gamma,n}[\mu(\lambda)]$$

and (2.6.52) together with the definition in (2.6.49) gives (2.6.51) and completes the proof of the theorem. $\qquad \square$

Formulas (2.6.49) and (2.6.50) combined are not sufficient to render explicit the dependence on n. To achieve this and obtain expansions similar to (2.6.48) for the polynomials $\Phi_{\gamma,n}$ we need one further step. This is provided by the following basic identity.

Proposition 2.13 *For* $\alpha \vdash m < n$ *let*

$$R_\alpha[x_1, x_2, \ldots, x_n; n] = \Xi_{\alpha(n)}[x(n)] \, , \qquad (2.6.53)$$

where for convenience we have set

$$x(n) = \big(x_1(n), x_2(n), \ldots, x_n(n)\big) \qquad (\text{with } x_i(n) = x_i + n - i).$$

Then

$$R_\alpha[x_1, x_2, \ldots, x_n; n]\Big|_{x_n=0} = R_\alpha[x_1, x_2, \ldots, x_{n-1}; n-1]. \qquad (2.6.54)$$

Proof Note that if $a_j \geq 1$ we may write

$$(y_i)_{a_j} = y_i (y_i - 1)_{a_j-1} \qquad (\text{for } i = 1, \ldots, n).$$

Thus

$$\det \left\| (y_i)_{a_j} \right\|_{i,j=1}^{n} \Big|_{\substack{y_n=0 \\ a_n=0}}$$

$$= y_1 y_2 \cdots y_{n-1} \det \begin{bmatrix} (y_1 - 1)_{a_1-1} & \cdots & (y_1 - 1)_{a_{n-1}-1} & 1 \\ (y_2 - 1)_{a_1-1} & \cdots & (y_2 - 1)_{a_{n-1}-1} & 1 \\ \vdots & \vdots & \vdots & \vdots \\ (y_{n-1} - 1)_{a_1-1} & \cdots & (y_{n-1} - 1)_{a_{n-1}-1} & 1 \\ 0 & \cdots & 0 & 1 \end{bmatrix}.$$

Thus from the definition in (2.6.19) we get that

$$\Xi_\alpha[y_1, y_2, \ldots, y_n]\Big|_{\substack{y_n=0 \\ a_n=0}} \qquad (2.6.55)$$

$$= \det \begin{bmatrix} (y_1 - 1)_{a_1-1} & \cdots & (y_1 - 1)_{a_{n-1}-1} \\ (y_2 - 1)_{a_1-1} & \cdots & (y_2 - 1)_{a_{n-1}-1} \\ \vdots & \cdots & \vdots \\ (y_{n-1} - 1)_{a_1-1} & \cdots & (y_{n-1} - 1)_{a_{n-1}-1} \end{bmatrix} \Bigg/ \prod_{1 \leq i < j \leq n-1} (y_i - y_j).$$

Note that for $l(\alpha) < n$ we have $\alpha_n(n) = 0$ and $\alpha_i(n) \geq 1$ for all $i < n$. Moreover, we also have

$$x_i(n) - 1 = x_i + n - i - 1 = x_i(n-1) \qquad (\text{for } i = 1, \ldots, n-1)$$

and

$$\alpha_i(n) - 1 = \alpha_i + n - i - 1 = \alpha_i(n-1) \qquad (\text{for } i = 1, \ldots, n-1).$$

Thus, setting $y_i = x_i(n)$ and $a_i = \alpha_i(n)$, in (2.6.55) immediately gives (2.6.54) precisely as asserted. □

Proposition 2.13 has the following remarkable corollary.

Proposition 2.14 *If $\alpha \vdash m$ then for any $n \geq m$ we have the power basis expansion*

$$\Xi_{\alpha(n)}(y_1, y_2, \ldots, y_n) = \sum_{|\rho| \leq m} c_\rho^\alpha(n)\, p_\rho(y_1, y_2, \ldots, y_n)\,. \qquad (2.6.56)$$

This given, define the coefficients $d_\rho^\alpha(n)$ through the equation

$$\sum_{|\rho| \leq m} c_\rho^\alpha(n)\, p_\rho \bigg|_{p_k \to R_k(n-1) + \sum_{s=1}^{k} \binom{k}{s}(n^s - (n-1)^s)q_s} = \sum_{|\rho| \leq m} d_\rho^\alpha(n)\, q_\rho \qquad (2.6.57)$$

where $q_1, q_2, \ldots, q_m$ are indeterminates and for $\rho = 1^{k_1} 2^{k_2} \cdots m^{k_m}$ we set

$$q_\rho = q_1^{k_1} q_2^{k_2} \cdots q_m^{k_m}\,.$$

Then for all $n \geq m$ we have

$$d_\rho^\alpha(n) = d_\rho^\alpha(m) \qquad \text{(for all } |\rho| \leq m\text{)}\,. \qquad (2.6.58)$$

Proof It is easily seen from the definition in (2.6.19) that Ξ_α is a polynomial of degree

$$a_1 + a_2 + \cdots + a_n - \binom{n}{2}$$

thus it follows that, when $\alpha \vdash m$, the polynomial in the left hand side of (2.6.56) is of degree m. Thus the power sum expansion of $\Xi_{\alpha(n)}(y_1, y_2, \ldots, y_n)$ must necessarily be of the form given in (2.6.56).

For convenience set for all $k \geq 1$

$$q_k(x_1, x_2, \ldots, x_n) = \sum_{i=1}^{n} \sum_{s=0}^{k} \binom{k}{s}(-i)^{k-a}\, R_a(x_i)\,. \qquad (2.6.59)$$

Note that by combining (2.6.27), (2.6.28), and (2.6.29) we derive that,

$$p_k(x_1(n), x_2(n), \ldots, x_n(n)) = R_k(n-1) + \sum_{s=1}^{k} \binom{k}{s}(n^s - (n-1)^s)q_s(x_1, x_2, \ldots, x_n)$$

$$(2.6.60)$$

Then it follows from the definition in (2.6.53) that

$$
R_\alpha[x_1, x_2, \ldots, x_n; n] = \sum_{|\rho| \le m} c_\rho^\alpha(n) \, p_\rho(x_1(n), x_2(n), \ldots, x_n(n))
$$

$$
\text{(by (2.6.57))} = \sum_{|\rho| \le m} d_\rho^\alpha(n) \, q_\rho(x_1, x_2, \ldots, x_n) \tag{2.6.61}
$$

where for $\rho = 1^{k_1} 2^{k_2} \cdots m^{k_m}$ we set

$$
q_\rho(x_1, x_2, \ldots, x_n) = q_1^{k_1}(x_1, x_2, \ldots, x_n) q_2^{k_2}(x_1, x_2, \ldots, x_n) \cdots q_m^{k_m}(x_1, x_2, \ldots, x_n) .
$$

Note next that Proposition 2.13 gives

$$
R_\alpha[x_1, x_2, \ldots, x_n; n]\Big|_{x_{m+1}=x_{m+2}=\cdots=x_n=0} = R_\alpha[x_1, x_2, \ldots, x_m; m]
$$

$$
\text{(by (2.6.61) for } m = n) = \sum_{|\rho| \le m} d_\rho^\alpha(m) \, q_\rho(x_1, x_2, \ldots, x_m) .
$$

$$\tag{2.6.62}$$

On the other hand (2.6.61) itself gives

$$
R_\alpha[x_1, x_2, \ldots, x_n; n]\Big|_{x_{m+1}=x_{m+2}=\cdots=x_n=0} \tag{2.6.63}
$$

$$
= \sum_{|\rho| \le m} d_\rho^\alpha(n) \, q_\rho(x_1, x_2, \ldots, x_n)\Big|_{x_{m+1}=x_{m+2}=\cdots=x_n=0} .
$$

But we plainly see from (2.6.59) that for all $k \ge 0$ we have

$$
q_k(x_1, x_2, \ldots, x_n)\Big|_{x_{m+1}=x_{m+2}=\cdots=x_n=0} = q_k(x_1, x_2, \ldots, x_m) . \tag{2.6.64}
$$

Combining (2.6.62), (2.6.63), and (2.6.64) gives

$$
\sum_{|\rho| \le m} d_\rho^\alpha(m) \, q_\rho(x_1, x_2, \ldots, x_m) = \sum_{|\rho| \le m} d_\rho^\alpha(n) \, q_\rho(x_1, x_2, \ldots, x_m)
$$

and this forces (2.6.58), completing the proof. □

Theorem 2.28 *Let γ be a partition of m all of whose parts are > 1 and set*

$$
W_\gamma(p_1, p_2, \ldots, p_m; n) = \frac{1}{z_\gamma} \sum_{\alpha \vdash m} \chi_\gamma^\alpha \sum_{|\rho| \le m} d_\rho^\alpha(m) \, p_\rho\Big|_{p_0 \to n} . \tag{2.6.65}
$$

Then for all $\lambda \vdash n$ *we have*

$$W_\gamma(p_1, p_2, \ldots, p_m ; n)\Big|_{p_k \to p_k[C(\lambda)]} = \omega_{\gamma,1^{n-m}} . \tag{2.6.66}$$

In particular the conjugacy class $C_{\gamma,1^{n-m}} \in CA(S_n)$ *is given by the formula*

$$C_{\gamma,1^{n-m}} = W_\gamma(p_1, p_2, \ldots, p_m ; n)\Big|_{p_k \to p_k[m_2,m_2,\ldots,m_n]} . \tag{2.6.67}$$

Proof It follows from Proposition 2.14 that, for $\gamma \vdash m$, the polynomial

$$\Phi_{\gamma,n} = \pi_n \Psi_{\gamma,n}$$

defined in (2.6.50), may also be written in the form

$$\Phi_{\gamma,n} = \frac{1}{z_\gamma} \sum_{\alpha \vdash m} \chi_\gamma^\alpha \sum_{|\rho| \leq m} d_\rho^\alpha(m) \, p_\rho .$$

We should note however that the definition (2.6.57) yields that some powers of q_0 will necessarily occur in the expansion on the right hand side of (2.6.57). Since for any $\lambda \vdash n$, we have $p_0[C(\lambda)] = n$, we see that in making the replacement $q_k \to p_k[C(\lambda)]$, the variable q_0 will necessarily be replaced by n. Thus (2.6.66) follows by combining Theorems 2.27 and 2.28. Formula (2.6.67) can then be derived from (2.6.66) the same way we derived (2.6.38) from (2.6.37). $\square$

The reader may find interesting to observe the remarkable simplicity of some of our polynomials W_γ given in the Appendix.

These notes would not be complete without the evaluation of the elementary symmetric functions at the Murphy elements. This can be stated as follows.

Theorem 2.29 *For* $s = 1, 2, \ldots, n$ *we have*

$$e_s(m_2, m_3, \ldots, m_n) = \sum_{l(\rho)=n-s} C_\rho . \tag{2.6.68}$$

Proof Remarkably, this identity is equivalent to a formula giving the principal specialization of Schur functions. This is shown by a purely combinatorial argument by Diaconis and Greene in [6]. We shall reverse the cart here and derive it from the Schur function identity. To be precise it is shown in [16, Ex. 4, p. 45] that for $\lambda \vdash n$ and for all $N \geq n$ we have

$$s_\lambda(x_1, \ldots x_N)\Big|_{x_1=\cdots=x_N=1} = \frac{1}{h_\lambda} \prod_{(i,j)\in\lambda} (N + j - i) . \tag{2.6.69}$$

On the other hand, Frobenius formula gives

$$s_\lambda(x_1, \ldots x_N) \Big|_{x_1=\cdots=x_N=1} \tag{2.6.70}$$

$$= \frac{\chi_\rho^\lambda}{z_\rho} p_\rho(x_1, \ldots x_N) \Big|_{x_1=\cdots=x_N=1}$$

$$= = \sum_{\rho \vdash n} \frac{\chi_\rho^\lambda}{z_\rho} N^{l(\rho)} .$$

Clearly the validity of (2.6.69) and (2.6.70) for all N implies the polynomial equality

$$\sum_{\rho \vdash n} \frac{\chi_\rho^\lambda}{z_\rho} t^{l(\rho)} = \frac{1}{h_\lambda} \prod_{(i,j)\in\lambda} (t + j - i) .$$

Thus it follows from Theorem 2.13 that

$$\prod_{k=2}^n (t + m_k) \, \chi^\lambda = \left(\sum_{\rho \vdash n} \frac{\chi_\rho^\lambda}{z_\rho} h_\lambda \, t^{l(\rho)} \right) \chi^\lambda$$

and equating coefficients of t^{n-s} we finally obtain

$$e_s(m_2, m_3, \ldots, m_n) \, \chi^\lambda = \left(\sum_{l(\rho)=n-s} \frac{\chi_\rho^\lambda}{z_\rho} h_\lambda \right) \chi^\lambda$$

$$= \left(\sum_{l(\rho)=n-s} c_\rho \right) \chi^\lambda$$

and the validity of this for all $\lambda \vdash n$ proves (2.6.68).   □

Appendix

$$\mathbf{W}_2 = \mathbf{p}_1$$

$$\mathbf{W}_3 = \frac{1}{2} n \, (n - 1) + \mathbf{p}_2$$

$$\mathbf{W}_4 = -(-3 + 2n) \, \mathbf{p}_1 + \mathbf{p}_3$$

$$\mathbf{W}_{22} = \frac{1}{2} n \, (n - 1) + \frac{1}{2} \mathbf{p}_1{}^2 - \frac{3}{2} \mathbf{p}_2$$

$$\mathbf{W}_5 = \frac{1}{6} n \, (n-1) \, (5 \, n - 19) + \mathbf{p}_4 - (-10 + 3 \, n) \, \mathbf{p}_2 - 2 \, \mathbf{p}_1{}^2$$

$$\mathbf{W}_{32} = -\frac{1}{2} \left(16 - 13 \, n + n^2\right) \mathbf{p}_1 - 4 \, \mathbf{p}_3 + \mathbf{p}_2 \, \mathbf{p}_1$$

$$\mathbf{W}_6 = 2 \, (3 \, n - 4) \, (n - 5) \, \mathbf{p}_1 + \mathbf{p}_5 - (-25 + 4 \, n) \, \mathbf{p}_3 - 6 \, \mathbf{p}_2 \, \mathbf{p}_1$$

$$\mathbf{W}_{42} = -\frac{4}{3} n \, (n-1) \, (2 \, n - 7) - 5 \, \mathbf{p}_4 + (-35 + 12 \, n) \, \mathbf{p}_2$$
$$+ \, \mathbf{p}_3 \, \mathbf{p}_1 - (-11 + 2 \, n) \, \mathbf{p}_1{}^2$$

$$\mathbf{W}_{33} = \frac{1}{8} n \, (n-1) \left(n^2 - 13 \, n + 34\right) - \frac{5}{2} \mathbf{p}_4 + \frac{1}{2} \mathbf{p}_2{}^2$$
$$- \frac{1}{2} \, (n-3) \, (n-10) \, \mathbf{p}_2 + 3 \mathbf{p}_1{}^2$$

$$\mathbf{W}_{222} = \frac{1}{2} \left(10 - 9 \, n + n^2\right) \mathbf{p}_1 + \frac{10}{3} \mathbf{p}_3 + \frac{1}{6} \mathbf{p}_1{}^3 - \frac{3}{2} \mathbf{p}_2 \, \mathbf{p}_1$$

$$\mathbf{W}_7 = -\frac{1}{24} n \, (n-1) \left(49 n^2 - 609 n + 1502\right) - \frac{9}{2} \mathbf{p}_2{}^2 - 8 \, \mathbf{p}_3 \mathbf{p}_1$$
$$+ \, 2 \, (-36 + 7 \, n) \, \mathbf{p}_1{}^2 + \frac{1}{2} \left(504 + 21 \, n^2 - 241 \, n\right) \mathbf{p}_2$$
$$- \frac{5}{2} \, (2 \, n - 21) \, \mathbf{p}_4 + \mathbf{p}_6$$

$$\mathbf{W}_{52} = 10 \, (-11 + 2 \, n) \, \mathbf{p}_3 + \frac{1}{6} \left(-174 \, n^2 + 889 \, n - 864 + 5 \, n^3\right) \mathbf{p}_1$$
$$- \, (3 \, n - 40) \, \mathbf{p}_2 \mathbf{p}_1 + \mathbf{p}_4 \mathbf{p}_1 - 6 \, \mathbf{p}_5 - 2 \, \mathbf{p}_1{}^3$$

$$\mathbf{W}_{43} = -\frac{1}{2} \, (n-5) \, (n-36) \, \mathbf{p}_3 + \frac{1}{2} \left(-240 - 53 \, n^2 + 251 \, n + 2 \, n^3\right) \mathbf{p}_1$$
$$- \, (-27 + 2 \, n) \, \mathbf{p}_2 \mathbf{p}_1 - 6 \, \mathbf{p}_5 + \mathbf{p}_3 \mathbf{p}_2$$

$$\mathbf{W}_{322} = -\frac{1}{4} \, (n-1) \left(n^2 - 25 \, n + 72\right) n + \frac{1}{2} \mathbf{p}_2 \mathbf{p}_1{}^2 - \frac{3}{2} \mathbf{p}_2{}^2 - 4 \, \mathbf{p}_3 \mathbf{p}_1$$
$$- \frac{1}{4} \left(n^2 - 25 \, n + 104\right) \mathbf{p}_1{}^2 + \frac{5}{4} \left(n^2 - 25 \, n + 60\right) \mathbf{p}_2 + 15 \, \mathbf{p}_4$$

$$\mathbf{W}_8 = \frac{1}{3} \left(48 \, n^2 + 3283 - 918 \, n\right) \mathbf{p}_3 - \frac{4}{3} \left(-264 \, n^2 - 945 + 1058 \, n + 16 \, n^3\right) \mathbf{p}_1$$
$$+ \, 2 \, (24 \, n - 197) \, \mathbf{p}_2 \mathbf{p}_1 - 2 \, (-49 + 3 \, n) \, \mathbf{p}_5 + \frac{32}{3} \mathbf{p}_1{}^3$$
$$- \, 12 \, \mathbf{p}_3 \mathbf{p}_2 + \mathbf{p}_7 - 10 \, \mathbf{p}_4 \mathbf{p}_1$$

$$\mathbf{W}_{62} = 3n(n-1)\left(3n^2 - 35n + 82\right) - 6\,\mathbf{p}_2\mathbf{p}_1{}^2 + 27\,\mathbf{p}_2{}^2 - (4n - 73)\,\mathbf{p}_3\mathbf{p}_1$$
$$+ \left(6n^2 - 110n + 379\right)\mathbf{p}_1{}^2 + \mathbf{p}_5\mathbf{p}_1 - \left(-564n + 1099 + 54\,n^2\right)\mathbf{p}_2$$
$$+ 10\,(-28 + 3n)\,\mathbf{p}_4 - 7\,\mathbf{p}_6$$

$$\mathbf{W}_{53} = -\frac{1}{24}n(n-1)\left(10n^3 - 273n^2 + 2243n - 4770\right) - 2\,\mathbf{p}_2\mathbf{p}_1{}^2$$
$$- \frac{1}{2}\,(6n - 65)\,\mathbf{p}_2{}^2 + \mathbf{p}_4\mathbf{p}_2 + 40\,\mathbf{p}_3\mathbf{p}_1 + \left(n^2 - 61n + 260\right)\mathbf{p}_1{}^2$$
$$+ \frac{7}{3}\left(-372 + n^3 + 206n - 27\,n^2\right)\mathbf{p}_2 - \frac{1}{2}\left(455 + n^2 - 61n\right)\mathbf{p}_4 - 7\,\mathbf{p}_6$$

$$\mathbf{W}_{44} = \frac{2}{3}n(n-1)\left(6n^2 - 62n + 139\right) + 9\,\mathbf{p}_2{}^2 + \frac{1}{2}\mathbf{p}_3{}^2 - (2n - 23)\,\mathbf{p}_3\mathbf{p}_1$$
$$+ \frac{1}{2}\left(4n^2 + 267 - 76n\right)\mathbf{p}_1{}^2 - \frac{3}{2}\,(8n - 21)\,(2n - 13)\,\mathbf{p}_2$$
$$+ 15\,(-7 + n)\,\mathbf{p}_4 - \frac{7}{2}\mathbf{p}_6$$

$$\mathbf{W}_{422} = \frac{1}{2}\left(560 - 121n + n^2\right)\mathbf{p}_3 - \frac{1}{6}\left(-471\,n^2 - 1890 + 22\,n^3 + 2069\,n\right)\mathbf{p}_1$$
$$+ \frac{1}{2}\,(30n - 247)\,\mathbf{p}_2\mathbf{p}_1 + 21\,\mathbf{p}_5 - \frac{1}{2}\,(-19 + 2n)\,\mathbf{p}_1{}^3 + \frac{1}{2}\mathbf{p}_3\mathbf{p}_1{}^2$$
$$- 5\,\mathbf{p}_4\mathbf{p}_1 - \frac{3}{2}\mathbf{p}_3\mathbf{p}_2$$

$$\mathbf{W}_{332} = \left(2n^2 - 62n + 245\right)\mathbf{p}_3 + \frac{1}{8}\left(2240 - 2490n + n^4 + 607\,n^2 - 38\,n^3\right)\mathbf{p}_1 - \frac{5}{2}\mathbf{p}_4\mathbf{p}_1$$
$$- \frac{1}{2}\left(-25n + n^2 + 190\right)\mathbf{p}_2\mathbf{p}_1 + \frac{1}{2}\mathbf{p}_1\mathbf{p}_2{}^2 + 21\,\mathbf{p}_5 + 3\,\mathbf{p}_1{}^3 - 4\,\mathbf{p}_3\mathbf{p}_2$$

$$\mathbf{W}_{2222} = \frac{1}{24}n(n-1)\left(3n^2 - 67n + 182\right) + \frac{1}{24}\mathbf{p}_1{}^4 - \frac{3}{4}\mathbf{p}_2\mathbf{p}_1{}^2 + \frac{9}{8}\mathbf{p}_2{}^2$$
$$+ \frac{10}{3}\mathbf{p}_3\mathbf{p}_1 + \frac{1}{4}\left(n^2 - 17n + 56\right)\mathbf{p}_1{}^2 - \frac{1}{4}\left(140 - 63n + 3\,n^2\right)\mathbf{p}_2 - \frac{35}{4}\mathbf{p}_4$$

Lecture 3
On Finite Dimensional $\mathfrak{sl}(2)$ Representations and an Application to Algebraic Combinatorics

3.1 Basic Identities

Throughout these notes **V** will be a finite dimensional K vector space and I will denote the identity operator. For all our applications there is no loss in specializing K to be the field of rational numbers, but at some point in our development we will need to assume that K is the field of complex numbers. We also assume that we have three linear operators E, F, H acting on V which satisfy the following three relations:

$$(a) \ [E, F] = H \ , \quad (b) \ [H, E] = 2E \ , \quad (c) \ [H, F] = -2F \ . \qquad (3.1.1)$$

As customary, the expression $L = (E + F + H)^*$ denotes the collection of all words in the alphabet $\{E, F, G\}$. Then the symbol $K[L]$ represents the collection of all finite linear combinations with coefficients in K of words of L. We intend to consider $K[L]$ as the subalgebra of linear operators on V generated by E, F and H. Our purpose here is to decompose the action of $K[L]$ on **V** into its irreducible constituents. To this end we need to derive a number of auxiliary identities. These will be stated as separate propositions.

Note first that if A, B, C are operators and we set

$$D_C A = [A, C] = AC - CA$$

then we can easily verify that

$$D_C AB = (D_C A) B + A (D_C B) . \qquad (3.1.2)$$

In other words D_C acts like *differentiation* on products of linear operators. We shall make systematic use of this fact in the sequel. However a word of caution

© The Editor(s) (if applicable) and The Author(s), under exclusive license
to Springer Nature Switzerland AG 2020
A. M. Garsia, Ö. Eğecioğlu, *Lectures in Algebraic Combinatorics*,
Lecture Notes in Mathematics 2277, https://doi.org/10.1007/978-3-030-58373-6_3

is necessary in regard to this notation. Certain identities satisfied by derivatives in ordinary calculus may not be valid for these differentiations. For instance familiar relation

$$D_C A^m = m A^{m-1} D_C A$$

can be used only when we know before hand that $D_C A$ commutes with A. A case in point is given by the proof of the following.

Proposition 3.1 *For any integer m we have*

$$(a) \quad H E^m = E^m (H + 2mI), \qquad\qquad (3.1.3)$$

$$(b) \quad H F^m = F^m (H - 2mI).$$

Proof Note we may write (3.1.3) (a) in the form

$$D_H E^m = -2m E^m .$$

Note further that for $m = 1$ this simply reduces to (3.1.1) (b). Thus we can proceed by induction and assume it true for m. This given, using (3.1.1) (b) and (3.1.2) we get

$$D_H E^{m+1} = (D_H E) E^m + E D_H E^m$$
$$= -2E^{m+1} - 2m E^{m+1} = -2(m + 1) E^{m+1}$$

as desired. A similar argument gives (3.1.3) (b).   □

As an immediate corollary we obtain the following remarkable fact.

Theorem 3.1 *The operators E and F are necessarily nilpotent.*

Proof Let E satisfy the equation

$$(E - \lambda I)^m \Gamma(E) = 0, \qquad\qquad (3.1.4)$$

where $m \geq 1$ and $\Gamma(t)$ is a polynomial that does not vanish at λ. There is no loss in assuming that $\Gamma(\lambda) = 1$. This given we can write

$$\Gamma(t) = 1 + (t - \lambda)Q(t) ,$$

from which we derive that

$$I = -(E - \lambda I)Q(E) + \Gamma(E) . \qquad\qquad (3.1.5)$$

Using (3.1.3) (a) we deduce that

$$D_H(E - \lambda I)^m = -2m\, E(E - \lambda I)^{m-1}$$

which may be rewritten as

$$-2m\, E(E - \lambda I)^{m-1} = (E - \lambda I)^m H - H(E - \lambda I)^m \ .$$

Multiplying on the right by $\Gamma(E)$ and using (3.1.4) gives

$$- 2m\, E(E - \lambda I)^{m-1}\Gamma(E) = (E - \lambda I)^m H\, \Gamma(E) \ . \tag{3.1.6}$$

On the other hand left multiplication of (3.1.5) by $-2m\, E(E - \lambda I)^{m-1}$ yields

$$-2m\, E(E - \lambda I)^{m-1} = 2m\, E(E - \lambda I)^m Q(E) - 2m\, E(E - \lambda I)^{m-1}\Gamma(E) \ .$$

Replacing the last term in this identity by the right hand side of (3.1.6) we get

$$-2m\, E(E - \lambda I)^{m-1} = 2m\, E(E - \lambda I)^m Q(E) + (E - \lambda I)^m H\, \Gamma(E) \ .$$

Since $\Gamma(E)$ commutes with any polynomial in E, this relation multiplied on the left by $\Gamma(E)$ may be written in the form

$$-2m E(E-\lambda I)^{m-1}\Gamma(E) = 2m E(E-\lambda I)^m\Gamma(E)Q(E)+(E-\lambda I)^m\Gamma(E)H\,\Gamma(E) \ ,$$

and from (3.1.4) we derive that

$$E(E - \lambda I)^{m-1}\Gamma(E) = 0.$$

But another use of (3.1.4) then yields that we must also have

$$\lambda(E - \lambda I)^{m-1}\Gamma(E) = E(E - \lambda I)^{m-1}\Gamma(E) = 0 \ .$$

If $\lambda \neq 0$ this process can be repeated and come to the final conclusion that (3.1.4) holds true only if

$$\Gamma(E) = 0 \ .$$

This shows that the minimal equation of E can only be $E^b = 0$ for some $b \leq dim\ \mathbf{V}$. A similar argument gives the nilpotency of F. $\qquad\qquad\square$

Proposition 3.2 *For any integer m we have*

$$(a) \quad F\, E^m = E^m F - m E^{m-1}(H + (m - 1)I) \tag{3.1.7}$$

$$(b) \quad E\, F^m = F^m E + m F^{m-1}(H - (m - 1)I) \ .$$

Proof We may write (3.1.7) (a) in the form

$$D_F E^m = m E^{m-1}(H + (m-1)I) .$$

Note that $m = 1$ gives $D_F E = H$ which is (3.1.1) (a). Thus we proceed by induction and assume (3.1.7) (a) true for m. We then get using (3.1.2) and the inductive hypothesis

$$
\begin{aligned}
D_F E^{m+1} &= (D_F E) E^m + E D_F E^m \\
&= H E^m + E(m E^{m-1}(H + (m-1)I) \\
\text{(by (3.1.3) (a))} = \ & E^m (H + 2mI) + m E^m H + m(m-1)E^m \\
&= E^m\big(((m+1)H + (m + m^2)I\big) = (m+1)E^m(H + mI)
\end{aligned}
$$

as desired. An analogous argument yields (3.1.7) (b).   □

This result enables us to identify the eigenvalues of H.

Theorem 3.2 *The eigenvalues of H are all integral and symmetrically distributed around the origin.*

Proof Let λ be an eigenvalue of H and let v be a corresponding eigenvector. If $\lambda = 0$ there is nothing to show. To be specific assume that $\lambda < 0$. This given, note that (3.1.3) (b) yields

$$H F^m v = F^m (Hv - 2mv) = -(2m - \lambda) F^m v . \tag{3.1.8}$$

Now let a be the smallest integer such that $F^{a+1}v = 0$. Note that, although Theorem 3.1 guarantees there must be such an integer, in this case there is an even simpler reason: (3.1.8) shows that the vectors $F^m v$ have distinct eigenvalues. Set then

$$u = F^a v$$

and note that we must have

$$(1) \ u \neq 0$$
$$(2) \ Fu = 0 \quad \text{and} \tag{3.1.9}$$
$$(3) \ H u = -k u$$

where for convenience we have set

$$k = 2a - \lambda . \tag{3.1.10}$$

Indeed, (1) and (2) follow from our choice of a and (3) is given by (3.1.8) for $m = a$. Equation (3.1.7) (a) then yields that for all m

$$F E^m u = E^m F u - m E^{m-1}(H + (m-1)I)u$$

$$\big(\text{by (2) and (3) of (3.1.9)}\big) = 0 - m E^{m-1}(-k + m - 1)u \qquad (3.1.11)$$

$$= m E^{m-1}(k + 1 - m)u \; .$$

Let now b be the smallest integer such that $E^{b+1}u = 0$. Using (3.1.11) with $m = b + 1$ yields

$$0 = F E^{b+1} u = (b+1)(k-b)E^b u \; .$$

Since by our choice of b the vector $E^b u$ cannot vanish we must necessarily have

$$2a - \lambda = k = b \geq 0 \; . \qquad (3.1.12)$$

This proves that λ is an integer as desired. To derive the last assertion note that (3.1.3) (a) and (3.1.9) (3) yield

$$H E^m u = (-k + 2m) E^m u \; . \qquad (3.1.13)$$

Moreover, $E^m u \neq 0$ for $0 \leq m \leq k(= b)$. This means that the integers

$$-k, \; -k+2, \; -k+4, \; \ldots, \; -4+k, \; -2+k, \; k$$

are all eigenvalues of H. Finally note that (3.1.13) for $m = a - \lambda$ gives (since $k = 2a - \lambda$)

$$H E^{a-\lambda} u = (-k + 2a - 2\lambda) E^{a-\lambda} u = -\lambda E^{a-\lambda} u \; . \qquad (3.1.14)$$

This proves that $-\lambda$ is also an eigenvalue of A, provided $E^{a-\lambda} \neq 0$ and this only requires that

$$a - \lambda \leq k \; .$$

But since $k = 2a - \lambda$ this inequality is

$$a - \lambda \leq 2a - \lambda$$

and this is trivially satisfied whatever the value of λ.

This completes our proof. □

The argument we used to prove Theorem 3.2 yields another important result.

Theorem 3.3 *Let v be an eigenvector of H with eigenvalue $-k \leq 0$. Suppose further that $Fv = 0$. Then the vectors*

$$v, Ev, E^2v, \ldots, E^kv \qquad (3.1.15)$$

are a basis of an irreducible $K[L]$-submodule $\mathbf{U}$ of $\mathbf{V}$.

Proof Our assumptions bring us to the case $a = 0$ of the previous argument. Thus $u = v$ in this case and the relations in (3.1.11) together with (3.1.9) (2) yield us that $\mathbf{U}$ is invariant under F. Likewise the fact that $E^{k+1}u = 0$ shows that $\mathbf{U}$ is invariant under E. Since (3.1.13) shows that each of the elements in (3.1.15) are eigenvectors of H we see that $\mathbf{U}$ is also invariant under H and the proof is complete. □

Remark 3.1 It is customary to call the collection in (3.1.15) an 𝔰𝔩(2) *string*. In this vein we will call here the vectors v and E^kv respectively the *head* and the *tail* of the string. Our goal is to show that V breaks up into a direct sum of strings. This done, to obtain a basis for V all we need to get is a basis for the subspace $sh(V)$ consisting of all *string heads*. That is

$$sh(V) = \bigoplus_{k \geq 0} \{v \in V \mid Hv = -kv \text{ and } Fv = 0\}.$$

3.2   Diagonalizability of H

Let us recall that an eigenvalue λ of an operator A is said to be *tame* if and only if

$$(A - \lambda I)^2 v = 0 \qquad \longrightarrow \qquad (A - \lambda I) v = 0 . \qquad (3.2.1)$$

Note that an operator all of whose eigenvalues are tame satisfies a polynomial equation with simple roots and must necessarily be diagonalizable. Our goal in this section is to put together the ingredients necessary to establish the diagonalizability of H. We start with the following.

Proposition 3.3 *If u is an eigenvector of H with eigenvalue $-k \leq 0$ and $Fu = 0$ then*

$$F^m E^m u = \frac{m!k!}{(k-m)!} u \qquad (\text{for } m = 0, 1, 2, \ldots, k) . \qquad (3.2.2)$$

Proof The identity is trivially true for $m = 0$. We shall thus inductively assume it true for m. This given, using (3.1.7) (a) (with $m \to m+1$) we get for $m < k$

$$F^{m+1} E^{m+1} u = F^m \left(FE^{m+1}u \right)$$

$$= F^m \left(E^{m+1}Fu - (m+1)(H - mI)E^m u \right)$$

$$= F^m\,(m+1)(k-m)E^m\,u$$

$$= (m+1)(k-m)F^m E^m\,u$$

$$\big(\text{by induction}\big) \;=\; (m+1)(k-m)\,\frac{m!k!}{(k-m)!}\,u$$

$$= \frac{(m+1)!k!}{(k-m-1)!}\,u\;.$$

This completes the induction and our proof.   □

Proposition 3.4 *For all integers m we have*

$$(a)\;\; (H-2I)^m\,E \;=\; EH^m \tag{3.2.3}$$

$$(b)\;\; (H+2I)^m\,F \;=\; FH^m$$

Proof The case $m=1$ of (3.2.3) (a) reduces to

$$(H-2I)E = EH$$

which is another way of writing (3.1.1) (b). Assuming inductively that (3.2.3) (a) is true for m we have

$$(H-2I)^{m+1}\,E \;=\; (H-2I)\big((H-2I)^m\,E\big)$$

$$= (H-2I)\big(EH^m\big) \;=\; EH^{m+1}\;.$$

This completes the induction and proves (3.2.3) (a). The other relation is established in the same manner.   □

Theorem 3.4 *The largest eigenvalue of H is tame.*

Proof Let k_0 be the largest eigenvalue of H. By symmetry $-k_0$ must be the smallest. Let $w \in \mathbf{V}$ satisfy

$$(H+k_0)^2\,w \;=\; 0\;. \tag{3.2.4}$$

Our goal is to show that

$$(H+k_0)\,w \;=\; 0.$$

We will show that the alternative

$$(H+k_0)\,w \;\neq\; 0 \tag{3.2.5}$$

leads to a contradiction. To this end let us set

$$u = (H + k_0)\, w \,.$$

Note first that (3.2.4) gives

$$H u = -k_0 u$$

and (3.1.3) (a) implies

$$H F u = -(k_0 + 2) F u \,.$$

To avoid contradicting the choice of k_0 we must then conclude that

$$F(H + k_0)\, w = F u = 0 \,. \tag{3.2.6}$$

Now note that

$$
\begin{aligned}
(H + 2I + k_0 I)^2 F &= (H + 2I)^2 F + 2k_0(H + 2I)F + k_0^2 F \\
\big(\text{by (3.2.3) (b)}\big) &= F\big(H^2 + 2k_0 H + k_0^2 I\big) \\
&= F(H + k_0 I)^2 \,.
\end{aligned}
$$

Thus (3.2.4) gives

$$(H + 2I + k_0 I)^2 F w = 0$$

but this may be rewritten as

$$H(H + 2I + k_0 I)F w = -(k_0 + 2)(H + 2I + k_0 I)F w$$

and, again, to avoid contradicting the choice of k_0 we must have

$$(H + 2I + k_0 I)F w = 0$$

but this is

$$H F w = -(2 + k_0) F w$$

and for the same reason we must conclude that

$$F w = 0 \,. \tag{3.2.7}$$

Note next that (3.1.3) (a) gives

$$H\, E^{k_0+1} = E^{k_0+1}(H + 2k_0 I + 2I)$$

and this may be rewritten as

$$(H - 2I - k_0 I)E^{k_0+1} = E^{k_0+1}(H + k_0 I).$$

Thus, a fortiori we must also have

$$(H - 2I - k_0 I)^2 E^{k_0+1} = E^{k_0+1}(H + k_0 I)^2.$$

But then (3.2.4) gives

$$(H - 2I - k_0 I)^2 E^{k_0+1} w = 0.$$

Now, here again, in order not to contradict the choice of k_0 we must accept that

$$(H - 2I - k_0 I)E^{k_0+1} w = 0$$

and for the same reason we are forced to accept that

$$E^{k_0+1} w = 0. \tag{3.2.8}$$

But this brings us to the final stretch of our proof. Indeed (3.2.8) gives

$$
\begin{aligned}
0 = F^{k_0+1} E^{k_0+1} w &= F^{k_0}\left(F E^{k_0+1} w \right) \\
\left(\text{by (3.1.7) (a)}\right) &= F^{k_0}\left(E^{k_0+1} F w - (k_0 + 1)E^{k_0}(H + k_0 I) \right) w \\
\left(\text{by (3.2.7)}\right) &= -(k_0 + 1)F^{k_0} E^{k_0}(H + k_0 I) w \\
\left(\text{by (3.2.2)}\right) &= -(k_0 + 1)(k_0!)^2\, (H + k_0 I) w\,.
\end{aligned}
$$

This contradicts (3.2.5) and completes our proof. □

Remark 3.2 To prove the tameness of all the other eigenvalues of H we need to separate the subspace generated by the action of $K[L]$ on the eigenvectors of eigenvalue $-k_0$ from the rest of $\mathbf{V}$. Our plan is to show that V may be broken up into the direct sum of two $K[L]$-invariant subspaces

$$\mathbf{V} = \mathbf{V}_1 \bigoplus \mathbf{V}_2 \tag{3.2.9}$$

where

(a) $\mathbf{V_1}$ contains all the eigenvectors of eigenvalue k_0,
(b) H is diagonalizable on $\mathbf{V_1}$,
(c) the largest eingenvalue of H in $\mathbf{V_2}$ is less than k_0.

This done, the diagonalizability of H immediately follows by an inductive argument which descends on the size of the largest eigenvalue of H.

Our first step in this program is given by the following basic result.

Theorem 3.5 *Let k_0 be the highest eigenvalue of H an let*

$$u_1, u_2, \ldots, u_N \tag{3.2.10}$$

be a basis for the eigenspace of H corresponding to the eigenvalue $-k_0$. Then the collection

$$\mathcal{B} = \bigcup_{i=1}^{N} \{u_i, Eu_i, \ldots, E^{k_0} u_i\} \tag{3.2.11}$$

forms an independent set and the linear span

$$\mathbf{V_1} = \mathcal{L}[\, E^m u_i \mid i = 1, 2, \ldots, N \; ; \; m = 0, 1, \ldots, k_0 \,] \; . \tag{3.2.12}$$

is a $K[L]$-invariant subspace of $\mathbf{V}$ which is a direct sum of irreducible $K[L]$-modules.

Proof Note that each u_i is killed by F since otherwise H would also have the eigenvalue $-k_0 - 2$. This given, from Theorem 3.3 it follows that the collection

$$\{u_i, Eu_i, \ldots, E^{k_0} u_i\}$$

spans an irreducible $K[L]$-module. Thus we need only show that $\mathcal{B}$ is an independent set. To this end suppose that for some constants $c_{i,m}$ we have

$$\sum_{m=0}^{k_0} \sum_{i=1}^{N} c_{m,i} \, E^m \, u_i = 0, \tag{3.2.13}$$

and note that the summands

$$\sum_{i=1}^{N} c_{0,i} \, u_i, \quad \sum_{i=1}^{N} c_{1,i} \, E \, u_i, \quad \sum_{i=1}^{N} c_{2,i} \, E^2 \, u_i \; , \ldots, \quad \sum_{i=1}^{N} c_{k_0,i} \, E^{k_0} \, u_i \; ,$$

are eigenvectors of H with respective eigenvalues

$$-k_0, \ -k_0 + 2, \ -k_0 + 4, \ \ldots, \ -k_0 + 2k_0 \ .$$

Thus (3.2.13) can hold true only if each of these summands separately vanishes. On the other hand if

$$\sum_{i=1}^{N} c_{m,i} \, E^m \, u_i \ = \ 0$$

then we must also have

$$\sum_{i=1}^{N} c_{m,i} \, F^m \, E^m \, u_i \ = \ 0 \, ,$$

but then (3.2.2) gives

$$\sum_{i=1}^{N} c_{m,i} \, u_i \ = \ 0$$

and the independence of the u_i's forces the vanishing of all the coefficients $c_{m,i}$, proving the independence of $\mathcal{B}$ as desired. $\qquad\square$

Note that since H acts diagonally on the basis $\mathcal{B}$, to complete our program we need only construct a $K[L]$-invariant complement $\mathbf{V_2}$ of $\mathbf{V_1}$. In fact, the construction of such an invariant complement would prove more than the diagonalizability of H, for it will also provide us with an algorithm for breaking up $\mathbf{V}$ into a direct sum of irreducible $K[L]$-modules.

It develops that the basic tools for carrying out the construction of $\mathbf{V_2}$ are supplied by *Casimir* operator

$$C \ = \ EF + FE + H^2/2 \ .$$

We shall present them here as three separate propositions.

Proposition 3.5 *C commutes with E, F and H.*

Proof Note that (3.1.1) (b), (c) give

$$D_H \, C \ = \ -2EF + 2EF + 2FE + -2FE + 0 = 0 \ .$$

Thus H and C commute. Using (3.1.1) (a) we get

$$D_E C = -EH + -HE + (2EH + 2HE)/2 = 0 .$$

Thus E and H commute as well. The result for F is shown in a similar manner.   □

Proposition 3.6 *If $Hu = -ku$ and $Fu = 0$ then for any m we have*

$$C E^m u = (k + k^2/2) E^m u \qquad (3.2.14)$$

Proof Note that, using (3.1.1) (a), we can rewrite C in the form

$$C = 2EF - H + H^2/2 . \qquad (3.2.15)$$

This gives

$$C u = 0 - Hu + H^2 u/2 = (k + k^2/2) u ,$$

and (3.2.14) follows immediately from the commutativity of C and E.   □

This result implies that every element of $\mathbf{V}_1$ is an eigenvector of C with eigenvalue $k_0 + k_0^2/2$. Setting for convenience

$$t_0 = k_0 + k_0^2/2 \qquad (3.2.16)$$

we may express this fact by writing

$$\mathbf{V}_1 \subseteq Nullspace[C - t_0 I] . \qquad (3.2.17)$$

We can convert this into an equality. In fact we can prove an even stronger result.

Proposition 3.7 *Let k_0 be the largest eigenvalue of H. Let $\mathbf{V}_1$ be defined as in (3.2.12) and t_0 be as in (3.2.16). Then for any $w \in \mathbf{V}$ we have*

$$(C - t_0 I)^p w = 0 \quad \longrightarrow \quad w \in \mathbf{V}_1 \qquad (3.2.18)$$

and in particular t_0 is a tame eigenvalue of C.

Proof Let a be the least integer such that

$$F^{a+1} w = 0 . \qquad (3.2.19)$$

Theorem 3.1 guarantees that such an integer exists. Our proof will proceed by induction on a. We start with $a = 0$. So let $Fw = 0$. This given, from (3.2.15) and (3.2.16) we derive that

$$
\begin{aligned}
(C - t_0)w &= (2EF - H + H^2/2 - k_0 - k_0^2/2)\, w \\
&= \left[-(H + k_0 I) + \frac{H^2 - k_0^2 I}{2} \right] w \qquad (3.2.20) \\
&= (H + k_0 I) \left(-1 + \frac{H - k_0 I}{2} \right) w \\
&= \frac{1}{2}(H + k_0 I)(H - (k_0 + 2)I)\, w \ .
\end{aligned}
$$

Thus the condition $(C - t_0 I)^p\, w = 0$ may be rewritten as

$$
\frac{1}{2^p}(H - (k_0 + 2)I)^p (H + k_0 I)^p\, w = 0 \ . \qquad (3.2.21)
$$

Now, the maximality of k_0 guarantees that the operator $H - (k_0 + 2)I$ is invertible. So (3.2.21) is equivalent to

$$
(H + k_0 I)^p\, w = 0 \ .
$$

However, since k_0 is a tame eigenvalue for H we see that we must also have

$$
(H + k_0 I)\, w = 0 \ .
$$

This places $w \in \mathbf{V}_1$ and completes the first step of the induction. So let (3.2.18) be true when $F^a\, w = 0$. Let $F^a w \neq 0$ and $F^{a+1} w = 0$. For convenience set

$$
u = F^a\, w \ . \qquad (3.2.22)
$$

The commutativity of C and H yields that

$$
(C - t_0)^p\, u = F^a\, (C - t_0 I)^p\, w = 0 .
$$

Since $Fu = 0$ we are reduced to the case just considered and u must also be an eigenvector of H with eigenvalue $-k_0$. In particular it must be a linear combination of the u_i's introduced in (3.2.10). So let

$$
u = \sum_{i=1}^{N} c_i\, u_i
$$

and set

$$\bar{w} = \frac{(k_0 - a)!}{a!k_0!} \sum_{i=1}^{N} c_i \, E^a \, u_i \ .$$

Since each u_i is killed by F we can apply (3.2.2) and deduce that

$$F^a \, \bar{w} = \sum_{i=1}^{N} c_i \, u_i \, = u = F^a w \ .$$

Now the difference $w - \bar{w}$ is killed by $(C - t_0 I)^p$ since w is so by hypothesis and $\bar{w}$ is even killed by $C - t_0 I$ since it lies in $\mathbf{V}_1$. Moreover as we have seen $F^a(w - \bar{w}) = 0$. Thus we can use the induction hypothesis and derive that $w - \bar{w} \in \mathbf{V}_1$. Since $w = \bar{w} + w - \bar{w}$ we have $w \in \mathbf{V}_1$ as desired. This completes the induction and our proof.   □

We can finally put all our pieces together and produce an invariant complement of $\mathbf{V}_1$. To this end note that since t_0 has been shown to be a tame eigenvalue of C, the minimal equation of C can be written in the form

$$(C - t_0 I) \, Q(C) = 0 \tag{3.2.23}$$

with $Q(t)$ a polynomial which does not vanish at t_0. There is no loss in assuming that $Q(t_0) = 1$. This gives

$$1 - Q(t) = (t - t_0)\Gamma(t) \ .$$

Thus we may write

$$I = Q(C) + (C - t_0 I) \, \Gamma(C) \ . \tag{3.2.24}$$

Theorem 3.6 *We have the direct sum decomposition*

$$\mathbf{V} = Q(C)\mathbf{V} \bigoplus (C - t_0 I)\Gamma(C)\mathbf{V} \ . \tag{3.2.25}$$

Since

$$\mathbf{V}_1 = Q(C)\mathbf{V} \tag{3.2.26}$$

and both summands are $K[L]$-invariant, we may take, in (3.2.9)

$$\mathbf{V}_2 = (C - t_0 I) \, \Gamma(C)\mathbf{V} \ .$$

Proof Note that the sum on the right hand side of (3.2.25) is direct since if we had two vectors v_1, v_2 such that

$$Q(C)\, v_1 = (C - t_0 I)\Gamma(C)\, v_2$$

then hitting this by $(C - t_0 I)$ and using (3.2.23) gives

$$(C - t_0 I)^2 \Gamma(C) v_2 = 0 \ .$$

But since t_0 is tame for C we must also have

$$(C - t_0 I)\Gamma(C) v_2 = 0 \ .$$

This implies that the summands in (3.2.25) share only the zero vector as desired. To show the equality in (3.2.25) we need only observe that (3.2.24) implies that every $w \in \mathbf{V}$ has the expansion

$$w = Q(C)w + (C - t_0 I)\, \Gamma(C)\, w\,. \tag{3.2.27}$$

Note next that since every element of $Q(C)\mathbf{V}$ is killed by $C - t_0 I$, Proposition 3.7 gives us that

$$Q(C)\mathbf{V} \subseteq \mathbf{V}_1 \ .$$

However we can reverse this inequality. In fact, since every element $v \in \mathbf{V}_1$ is killed by $C - t_0 I$, the expansion in (3.2.27) yields that it may be written in the form

$$v = Q(C)\, v \ .$$

Thus

$$\mathbf{V}_1 \subseteq Q(C)\mathbf{V} \ .$$

This proves (3.2.26). $\qquad\qquad\qquad\qquad\qquad\qquad\qquad\qquad\qquad\qquad\qquad$ □

Since the $K[L]$-invariance of the space $(C - t_0 I)\Gamma(C)\mathbf{V}$ is an immediate consequence of the commutativity of C with E, F and H we see that (3.2.25) is precisely the decomposition of $\mathbf{V}$ we were looking for and our program for establishing the diagonalizability of H is now complete.

There are two further results concerning the eigenvectors of H that are worth including here. To begin we have

Proposition 3.8 *Let $Hu = -ku$ with $k \geq 0$ and let*

$$(a)\ F^a u \neq 0 \qquad and \qquad (b)\ F^{a+1}u = 0 \tag{3.2.28}$$

Then there exists a unique set of vectors $u_0, u_1, \ldots, u_a$ *such that*

$$u_0 + E u_1 + \cdots + E^a u_a = u \tag{3.2.29}$$

with

(a) $F u_i = 0$ *and* (b) $H u_i = -(k + 2i)u$ *(for i = 0, 1, \ldots, (a)*.

$$\tag{3.2.30}$$

Proof Say $a = 0$. Then we can take $u_0 = u$ and we are done. So we may proceed by induction. Suppose the result true up to $a - 1 \geq 0$. Let $F^a u \neq 0$ and $F^{a+1}u = 0$. Now set

$$\bar{u} = c E^a F^a u. \quad \left(\text{with } c = \frac{(k+a)!}{(k+2a)!a!} \right). \tag{3.2.31}$$

Note that (3.1.3) (b) with $m = a$ gives

$$H F^a u = F^a (H - 2aI)u = F^a(-k - 2a)u = -(k + 2a) F^a u. \tag{3.2.32}$$

This given, applying formula (3.2.2) with $m = a$ to (3.2.29) we get

$$F^a \bar{u} = c F^a E^a \left(F^a u \right) = c \frac{(k+2a)!a!}{(k+2a-a)!} F^a u = F^a u. \tag{3.2.33}$$

Since the vector

$$v = u - \bar{u} \tag{3.2.34}$$

satisfies the hypothesis $F^a v = 0$, by induction we may write

$$v = u_0 + u_1 + \cdots + u_{a-1}$$

with $F u_i = 0$, for $i = 0, 1, \ldots a - 1$. Combining this with (3.2.34) we get

$$u = u_0 + E u_1 + \cdots + E^{a-1} u_{a-1} + \bar{u}$$
$$= u_0 + E u_1 + \cdots + E^{a-1} u_{a-1} + c E^a \left(F^a u \right).$$

This may be rewritten as

$$u = u_0 + E u_1 + \cdots + E^{a-1} u_{a-1} + E^a u_a \tag{3.2.35}$$

with

$$u_a = c F^a u.$$

Since by assumption $F u_a = c F^{a+1} u = 0$ we see that (3.2.35) completes the induction and establishes existence. Finally note that under conditions (a) and (b) of (3.2.30) the summand $E^i u_i$ in (3.2.29) is an eigenvector of the Casimir operator C with eigenvalue $(k + 2i) + (k + 2i)/2$. Thus the sum in (3.2.29) can never vanish. This proves uniqueness and completes our proof. $\qquad\square$

Proposition 3.8 has the following important consequence.

Theorem 3.7 *Let $Hu = -ku$ with $k \geq 0$, suppose that $Fu = 0$ then we cannot have*

$$u = E v . \tag{3.2.36}$$

Proof Suppose such a v exists giving (3.2.36). Then let a be such that

$$F^a v \neq 0 \quad \text{and} \quad F^{a+1} v = 0 .$$

Then from the previous proposition we derive that

$$v = v_0 + E v_1 + \cdots + E^a v_a \quad \text{(with } F v_i = 0) \tag{3.2.37}$$

with

$$(a) \ F v_i = 0 \quad \text{and} \quad (b) \ H v_i = -(k + 2i) v_i . \tag{3.2.38}$$

In particular (3.2.2) gives

$$F^i E^i u_i = \frac{i!(k + 2i)!}{(k + i)!} u_i . \tag{3.2.39}$$

Now combining (3.2.36) and (3.2.37) we get

$$u = E v = E v_0 + E^2 v_1 + \cdots + E^{a+1} v_a$$

and in particular, $Fu = 0$ gives

$$0 = F E v_0 + F E^2 v_1 + \cdots + F E^{a+1} v_a . \tag{3.2.40}$$

Since (3.2.39) and (3.2.38) (a) imply that

$$F^j E^i v_i = 0 \quad \text{(for all } j > i) ,$$

applying F^a to (3.2.40) and using formula (3.2.39) with $m = a + 1$ we derive that

$$0 = F^{a+1} E^{a+1} v_a = \frac{(a + 1)!(k + 2a + 2)!}{(k + a + 1)!} v_a .$$

This gives $v_a = 0$. We can repeat this argument, successively multiplying (3.2.35) by $F^{a-1}, F^{a-2}, \ldots$, and obtain that all the v_i must necessarily vanish there by reaching a contradiction. This completes our proof.   □

For applications of $\mathfrak{sl}(2)$ theory it will be convenient to study the action of E and F on the eigenspaces of H. To this end let k_0 as before denote the highest eigenvalue of H and set

$$(a) \; V_k = \{v \in V \mid Hv = k\,v\} \quad (\text{for } -k_0 \le k \le k_0)$$

$$(b) \; U_k = \{v \in V \mid Hv = -k\,v \text{ and } F\,v = 0\} \quad (\text{for } 0 \le k \le k_0).$$

With this, we have the following fundamental fact.

Theorem 3.8 *a For any $1 \le k \le k_0$, the operator E^k, yields a nonsingular map of V_{-k} onto V_k and F^k yields a nonsingular map of V_k onto V_{-k}. In particular it follows that*

$$\dim V_{-k} = \dim V_k \qquad (\text{for all } 1 \le k \le k_0). \tag{3.2.41}$$

Proof We will only show that for all $v \in V_{-k}$

$$E^k v = 0 \implies v = 0 \tag{3.2.42}$$

since the corresponding result for F is proved in an entirely analogous manner, by inverting the roles of E and F.

Now Proposition 3.8 yields that if $v \in V_{-k}$ then for some $a \ge 0$ we can write v in the form

$$v = v_0 + E\,v_1 + \cdots + E^a v_a \tag{3.2.43}$$

with

$$(a) \; F v_i = 0 \text{ and } (b) \; Hv_i = -(k+2i)\,v_i \quad (\text{for } i = 0, 1, \ldots, (a)). \tag{3.2.44}$$

Note that if $a = 0$ then (a) and (b) for $i = 0$ give $Fv = 0$ and $Hv = -kv$ thus from Proposition 3.3 we derive that

$$F^k E^k v = k!^2\, v$$

and this proves (3.2.42) in this case. So we may proceed by induction on a. To this end note first that using (a) and (b) of (3.2.44) together with Proposition 3.3 we derive that

$$F^m E^m v_i = \frac{m!(k+2i)!}{(k+2i-m)!}\, v_i \quad (\text{for all } 0 \le m \le k+2i). \tag{3.2.45}$$

In particular we have

$$F^i E^i v_i = \frac{i!(k+2i)!}{(k+i)!} v_i . \qquad (3.2.46)$$

Now if $E^k v = 0$ from the expansion in (3.2.43) we derive that

$$0 = E^k v_0 + \cdots E^{k+i} v_i + \cdots + E^{k+a} v_a$$

and a fortiori we must also have

$$0 = F^{k+a} E^k v_0 + \cdots + F^{k+a} E^{k+i} v_i + \cdots + F^{k+a} E^{k+a} v_a . \qquad (3.2.47)$$

Now (3.2.45) for $m = k + i$ gives

$$F^{k+i} E^{k+i} v_i = \frac{(k+i)!(k+2i)!}{i!} v_i \qquad (3.2.48)$$

and (3.2.44) (a) gives

$$F^{k+a} E^{k+i} v_i = 0 \qquad (\text{for all } 0 \le i < (a)) .$$

Thus using (3.2.48) with $i = a$ from (3.2.47) we derive that

$$0 = F^{k+a} E^{k+a} v_a = \frac{a!(k+2a)!}{(k+a)!} v_a .$$

This forces $v_a = 0$ and the inductive hypothesis gives $u = 0$ completing the proof of (3.2.42). □

This brings us to a result which has been used in various contexts to prove unimodality.

Theorem 3.9 *For every* $2 \le k \le k_0$

(i) *the mapping from* V_{-k} *into* V_{-k+2} *induced by E is injective.*
(ii) *the mapping from* V_{-k+2} *into* V_{-k} *induced by F is onto and thus*

$$\dim U_k = \dim V_{-k} - \dim V_{-k-2} . \qquad (3.2.49)$$

Analogously we also have

(i') *the mapping from* V_k *into* V_{k-2} *induced by F is injective.*
(ii') *the mapping from* V_{k+2} *into* V_k *induced by F is onto.*

In particular (3.2.49) and (3.2.41) yield us the inequalities

$$\dim V_{-k} \le \dim V_{-k+2} \le \cdots \le \dim V_{-2} \le \dim V_0 \qquad (3.2.50)$$
$$\dim V_0 \ge \dim V_2 \ge \cdots \ge \dim V_{k-2} \ge \dim V_k$$

for all even $k \le k_0$, and

$$\dim V_{-k} \le \dim V_{-k+2} \le \cdots \le \dim V_{-3} \le \dim V_{-1} \qquad (3.2.51)$$
$$\dim V_1 \ge \dim V_3 \ge \cdots \ge \dim V_{k-2} \ge \dim V_k$$

for all odd $k \le k_0$.

Proof For $u \in V_{-k}$ we cannot have $Eu = 0$ since that would give $E^k u = 0$ and contradict Theorem 3.8. Now for $u \in V_{-k-2}$ Proposition 3.8 gives us the expansion

$$u = u_0 + E u_1 + \cdots + E^a u_a \qquad (3.2.52)$$

with the u_i satisfying (a) and (b) of (3.2.30) with k replaced by $k + 2$. In particular from (3.1.7) (a) we derive that

$$F E^{i+1} u_i = (i+1)(k+2+i) E^i u_i . \qquad (3.2.53)$$

Thus if we set

$$v = \sum_{i=0}^{a} \frac{E^i u_i}{(i+1)(k+2+i)}$$

then (3.2.52) and (3.2.53) give

$$u = F v . \qquad (3.2.54)$$

Since from (3.1.3) (a) and (3.2.30) (b) we easily derive that $v \in V_{-k}$, the equality in (3.2.54) proves that F maps V_{-k} onto V_{-k-2}. In particular this gives the equality

$$\dim_{\text{Range } F} V_{-k} = \dim V_{-k-2}$$

and (3.2.49) immediately follows from the simple identity

$$\dim V_{-k} = \dim_{\text{Range } F} V_{-k} + \dim_{\text{Ker } F} V_{-k} .$$

Our next two results bring to light the action of E, F and H on the vector space spanned by the elements of an $\mathfrak{sl}(2)$ string. To begin we have

Theorem 3.10 *If $F u = 0$ and $H u = -m u$ and we set*

$$e_j = (m - j)! \, E^j u \tag{3.2.55}$$

then the action of E, F, H on the basis $\langle e_0, e_1, \ldots, e_m \rangle$ is given by the matrices

(a) $\tilde{E} = \left\| (m - j)\chi(i = j + 1) \right\|_{i,j=0}^{m}$

(b) $\tilde{F} = \left\| j \, \chi(i = j - 1) \right\|_{i,j=0}^{m}$ $\qquad (3.2.56)$

(c) $\tilde{H} = \left\| (2j - m)\chi(i = j) \right\|_{i,j=0}^{m}$.

Proof Recall that we make the convention that the action of an operator T on a basis is to be expressed by right multiplication by a matrix. Thus to prove (3.2.56) (a) we need to show that the matrix $A = \left\| a_{ij} \right\|_{i,j=0}^{m}$ determined by the condition

$$E \langle e_0, e_1, \ldots, e_m \rangle = \langle e_0, e_1, \ldots, e_m \rangle A \tag{3.2.57}$$

is given by the right hand side of (3.2.56) (a).

Note that the definition in (3.2.55) gives

$$E \, e_j = (m - j)! \, E^i u = (m - j) \, e_{j+1} . \tag{3.2.58}$$

Now, by equating columns, we see that (3.2.57) is equivalent to the relations

$$E \, e_j = \sum_{i=0}^{m} e_i \, a_{ij} . \tag{3.2.59}$$

Equating the right hand sides of (3.2.58) and (3.2.59) we see that we must have

$$a_{ij} = \begin{cases} 0 & \text{if } i \neq j + 1 \\ m - j & \text{if } i = j + 1 \end{cases}$$

and this is simply another way of writing (3.2.56) (a).

The same mechanism yields (3.2.56) (b) and (c). Except that for (3.2.56) (b) we first use (3.1.7) (a) for $m = j$ to get

$$F E^j u = -(j) E^{j-1} (-m + j - 1) u = j(m - j + 1) \, E^{j-1} u$$

or, equivalently (by (3.2.55))

$$F e_j = (m - j)! F E^j u = j(m + 1 - j)! E^{j-1} u = j e_{j-1}.$$

The remaining steps are is as in the previous case. For (3.2.56) (c) we start with (3.1.3) (a), for $m = j$ and get

$$H E^j u = (-m + 2j) E^j u$$

or equivalently

$$H e_j = (2j - m) e_j$$

and the remaining steps are is as in the previous cases.   □

To proceed we need a result which extends the identity in (3.2.2).

Proposition 3.9 *If $Hu = -m u$ and $Fu = 0$ then*

$$F^j E^s u = \begin{cases} 0 & \text{if } s < j \\ \dfrac{s!(m - s + j)!}{(s - j)!(m - s)!} E^{s-j} u & \text{if } s \geq j . \end{cases} \tag{3.2.60}$$

Proof Formula (3.1.7) (a) gives

$$F E^s u = s(m - s + 1) E^{s-1} u . \tag{3.2.61}$$

Since $Fu = 0$ the case $j = 1$ of (3.2.60) holds true. So we may proceed by induction on j and assume (3.2.60) true for j. Then (3.2.60) immediately gives

$$F^{j+1} E^s u = 0 \qquad (\text{if } s \leq j)$$

while for $s \geq j + 1$ we get

$$F^{j+1} E^s u = \frac{s!(m - s + j)!}{(s - j)!(m - s)!} F E^{s-j} u$$

$$(\text{by } (3.2.61) \text{ for } s \to s - j) = \frac{s!(m - s + j)!}{(s - j)!(m - s)!} (s - j)(m - s + j + 1) E^{s-j-1} u$$

$$= \frac{s!(m - s + j + 1)!}{(s - j - 1)!(m - s)!} E^{s-j-1} u .$$

This completes the induction and the proof.   □

Theorem 3.11 *If $F u = 0$ and $H u = -m u$ then the operator*

$$\frac{(m-j)! i!}{j! m!^2 (m-i)!} F^{m-i} E^m F^j \tag{3.2.62}$$

acts on the basis

$$\langle u, E u, \ldots, E^m u \rangle$$

by the matrix

$$U^{(ij)} = \left\| \chi(r=i) \chi(s=j) \right\|_{r,s=0}^{m}$$

and consequently the operator

$$\sum_{i,j=0}^{m} a_{ij} \frac{(m-j)! i!}{j! m!^2 (m-i)!} E^{m-i} E^m F^j \tag{3.2.63}$$

acts by the matrix $A = \left\| a_{ij} \right\|_{i,j=0}^{m}$. In the same vein, if we set

$$e_j = (m-j)! E^j u \quad \text{(for } j = 0, \ldots, m)$$

then the operators

$$\frac{i!}{j! m!^2} F^{m-i} E^m F^j \quad \text{and} \quad \sum_{i,j=1}^{m+1} a_{ij} \frac{i!}{j! m!^2} F^{m-i} E^m F^j$$

act on the basis $\langle e_0, e_1, \ldots, e_m \rangle$ likewise by $U^{(ij)}$ and A respectively.

Proof From (3.2.60) it follows that

$$E^m F^j E^s u = \begin{cases} 0 & \text{if } s < j, \\ \dfrac{s! (m-s+j)!}{(s-j)! (m-s)!} E^m E^{s-j} u & \text{if } s \geq j. \end{cases}$$

But

$$E^m E^{s-j} u = 0 \quad \text{for all } s > j.$$

Thus

$$E^m F^j E^s u = \begin{cases} 0 & \text{if } s \neq j, \\ \dfrac{j!m!}{(m-j)!} E^m u & \text{if } s = j. \end{cases}$$

In particular

$$F^{m-i} E^m F^j E^s u = \begin{cases} 0 & \text{if } s \neq j \\ \dfrac{j!m!}{(m-j)!} F^{m-i} E^m u & \text{if } s = j \end{cases} \tag{3.2.64}$$

and (3.2.61) for $j \rightarrow m - i$ and $s \rightarrow m$ gives

$$F^{m-i} E^m u = \frac{m!(m-i)!}{i!} E^i u$$

so that (3.2.64) becomes

$$F^{m-i} E^m F^j E^s u = \begin{cases} 0 & \text{if } s \neq j \\ \dfrac{j!\, m!^2 (m-i)!}{(m-j)!\, i!} E^i u & \text{if } s = j. \end{cases}$$

In other words

$$\frac{(m-j)!\, i!}{j!\, m!^2 (m-i)!} F^{m-i} E^m F^j E^s u = \begin{cases} 0 & \text{if } s \neq j \\ E^i u & \text{if } s = j. \end{cases}$$

This proves that the operator in (3.2.62) acts by $U^{(ij)}$ and consequently the operator in (3.2.63) acts by A. The remaining assertions are immediate consequences of the definition of the basis $\langle e_0, e_1, \ldots, e_m \rangle$. □

N. Wallach pointed out to us the following remarkable consequence of Theorem 3.11.

Theorem 3.12 *Let $\mathcal{M}_m$ and $\mathcal{M}_n$ respectively denote the spaces of $m \times m$ and $n \times n$ matrices over $\mathbf{Q}$ and let $\phi : \mathcal{M}_m \rightarrow \mathcal{M}_n$ be a map satisfying, for all $A, B \in \mathcal{M}_m$ and $c \in \mathbf{Q}$:*

$$(a)\ \phi(A + B) = \phi(A) + \phi(B)$$

$$(b)\ \phi(AB) = \phi(A)\phi(B) \tag{3.2.65}$$

$$(c)\ \phi(c\,A) = c\,\phi(A)$$

Then $n = Nm$ and $\phi(A)$ (up to a similarity) is none other than a block diagonal matrix consisting of N copies of A. More precisely we have a nonsingular $mN \times mN$ matrix D yielding the decomposition

$$\phi(A) = D\left(\bigoplus_{i=1}^{N} A\right) D^{-1}. \tag{3.2.66}$$

Proof Let $\tilde{E}, \tilde{F}, \tilde{H}$ be the matrices defined in (3.2.56). Let V_n be the n-dimensional vector space over $\mathbf{Q}$ and let E, F and H be the linear operators on V_n defined by setting for each $v \in V_n$

$$E v = \phi(\tilde{E}) v, \quad F v = \phi(\tilde{F}) v, \quad H.v = \phi(\tilde{H}) v,$$

Property (3.2.65) (b) assures that E, F and H satisfy the basic relations in (3.1.1). This immediately yields that V_n decomposes into a direct sum of $\mathfrak{sl}(2)$-strings

$$V_n = \bigoplus_{i=1}^{N} \mathcal{L}[u_i, Eu_i, \dots, E^{k_i} u_i] \tag{3.2.67}$$

with

$$(a)\ H u_i = -k_i u_i, \quad (b)\ F u_i = 0. \tag{3.2.68}$$

However, note that the definitions in (3.2.56) assure that the Casimir matrix

$$\tilde{C} = 2\tilde{E}\tilde{F} - \tilde{H} + \tilde{H}^2/2$$

reduces to the multiple $(m + m^2/2)\tilde{I}$ of the $m \times m$ identity matrix $\tilde{I}$. But then, the identities in (3.2.65) yield that the Casimir operator

$$C = 2EF - H + H^2/2$$

must also be of the form

$$C = (m + m^2/2)I$$

with I the identity operator on V_n. On the other hand, by Proposition 3.6, the relations in (3.2.68) yield the equality

$$C u_i = (k_i + k_i^2/2)u_i$$

and this forces the equality

$$k_i + k_i^2/2 = m + m^2/2. \tag{3.2.69}$$

But for $k_i \geq 0$ this can only hold true for

$$k_i = m .$$

Since (3.2.69) must hold for all i the decomposition in (3.2.67) must reduce to

$$V_n = \bigoplus_{i=1}^{N} \mathcal{L}[u_i, E u_i, \ldots, E^m u_i] . \tag{3.2.70}$$

with

$$(a) \ H u_i = -m \, u_i , \quad (b) \ F u_i = 0 . \tag{3.2.71}$$

This proves $n = Nm$. We may thus introduce the basis

$$\mathcal{B} = \bigcup_{i=1}^{n} \{ e_0^{(i)}, e_1^{(i)}, \ldots, e_m^{(i)} \} \tag{3.2.72}$$

with

$$e_s^{(i)} = (m - i)! \, E^s \, u^{(i)}$$

and conclude from Theorem 3.12 that the operator

$$T_A = \sum_{i,j=0}^{m} a_{ij} \frac{i!}{j! m!^2} F^{m-i} E^m F^j$$

acts on each basis block

$$\langle e_0^{(i)}, e_1^{(i)}, \ldots, e_m^{(i)} \rangle$$

by the matrix $A = \|a_{ij}\|_{i,j=0}^{m}$. However our definition of E, F, H together with the relation in (3.2.65) gives us that, for any $v \in V_n$ we have

$$T_A v = \phi \left(\sum_{i,j=0}^{m} a_{ij} \frac{i!}{j! m!^2} \tilde{F}^{m-i} \tilde{E}^m \tilde{F}^j \right) v .$$

But again from Theorem 3.11 we easily derive that

$$\sum_{i,j=0}^{m} a_{ij} \frac{i!}{j! m!^2} \tilde{F}^{m-i} \tilde{E}^m \tilde{F}^j = A .$$

This gives that $\phi(A)$ acts by A on each basis block proving the decomposition in (3.2.66), with D the matrix relating the standard basis of V_n with the basis $\mathcal{B}$ in (3.2.72). This completes our argument. □

It is remarkable that this important theorem in representation theory [12] can be proved as an application of $\mathfrak{sl}(2)$ theory.

3.3 Spernerity of $L[m, n]$

We recall that a finite *poset* (partially ordered set) is a pair $P = (\Omega, \leq)$ consisting of a finite set Ω together with a relation "$<$" satisfying the three conditions

(i) $x \leq x$ for all $x \in \Omega$
(ii) $x \leq y$ $y \leq x$ $\implies$ $x = y$
(iii) $x \leq y$ $y \leq z$ $\implies$ $x \leq z$

If $x \leq y$ and $x \neq y$ we write $x < y$. If $x < y$ and there is no element $z \in \Omega$ satisfying $x < z < y$ we write $x \rightarrow y$ and say that x *is a predecessor of* y or that y *is a successor of* x. We may also say that y *covers* x. When one of $x < y$ or $y < x$ holds true we say that x and y are *comparable* otherwise x and y are called *incomparable*.

The *Hasse diagram of* P is Ω drawn with a line, sometimes with an arrow, joining each element to its successor. An element $x \in \Omega$ is said to be *minimal* if x has no predecessor in P and *maximal* if it has no successor. A *chain* of P is a subset $C \subseteq \Omega$, totally ordered by $<$. That is we have

$$C = \{x_0, x_1, x_2, \ldots, x_k\} \quad \text{with} \quad \{x_0 < x_1 < x_2 < \cdots < x_k\}$$

In this case we say that C *joins* x_0 to x_k and k is called the *length* of C, the latter is usually denoted $l(C)$. By contrast we say that a subset $A \subseteq \Omega$ is an *antichain* if all pairs of elements of A are incomparable.

A chain C is called *unrefinable* or *saturated* if and only if

$$C = \{x_0, x_1, x_2, \ldots, x_k\} \quad \text{with} \quad \{x_0 \rightarrow x_1 \rightarrow x_2 \rightarrow \cdots \rightarrow x_k\} .$$

A *maximal chain* is a saturated chain which is not a proper subset of another chain. A poset whose maximal chains have all the same length is called *ranked*. If P is ranked then it is convenient to add to P a *zero* element denoted $\hat{0}$ to be a predecessor of all minimal elements of P. Likewise if P has more than one maximal element it is convenient to add to P a *unit* element denoted $\hat{1}$ to be a successor of all maximal elements of P. This done it follows that all unrefinable chains joining $\hat{0}$ to a given element x have the same length. This common length is called the *rank of* x and is denoted $r(x)$. The rank of $\hat{1}$ is called the *rank of* P and is denoted $r(P)$.

The polynomial

$$F_P(q) = \sum_{x \in \Omega} q^{r(x)} \tag{3.3.1}$$

is called the *rank generating function of* P. The collection of elements of P of a given rank is called *a rank row of* P. It will be convenient to set

$$\mathcal{R}_k(P) = \{x \in \Omega \mid r(x) = k\} \tag{3.3.2}$$

and refer to it as the kth *rank row of* P. The cardinality of $\mathcal{R}_k(P)$ will be denoted $r_k(P)$. In particular, if $r(P) = k_0$ the definition in (3.3.1) may be written in the form

$$F_P(q) = \sum_{k=0}^{k_0} r_k(P) q^k . \tag{3.3.3}$$

We say that P is *rank symmetric* if and only if

$$r_{k_0-k}(P) = r_k(P) \qquad \text{(for all } 0 \le k \le k_0). \tag{3.3.4}$$

In view of (3.3.3) we see that this condition is equivalent to the identity

$$F_P(q) = q^{r(P)} F_P(1/q) . \tag{3.3.5}$$

The poset P is said to be *rank unimodal* if the sequence $r_k(P)$ weakly increases from 1 to its maximum and then weakly decreases back to 1. More precisely, for some $0 < k_1 < k_0$ we have

$$1 \le r_1(P) \le r_2(P) \le \cdots \le r_{k_1-1}(P) \le r_{k_1}(P) \quad \text{and} \tag{3.3.6}$$
$$r_{k_1}(P) \ge r_{k_1+1}(P) \ge \cdots \ge r_{k_0-1}(P) \ge 1 .$$

Note that when $P = (2^{[n]}, \subseteq)$ is the lattice of subsets of the n-set $[n] = \{1, 2, \ldots, n\}$ and the rank of a subset $S \subseteq [n]$ is given by its size $|S|$, we have

$$F_P(q) = (1+q)^n = \sum_{k=0}^{n} \binom{n}{k} q^k . \tag{3.3.7}$$

In this case the relations in (3.3.6) are easily shown to be true with $k_1 = [\frac{n}{2}]$. Moreover the symmetry of binomial coefficients yields the relations in (3.3.4) for $k_0 = n$ as well. That is the lattice of subsets $(2^{[n]}, \subseteq)$ is rank symmetric and rank unimodal. It is clear from the definition of rank that

$$x < y \rightarrow r(x) < r(y)$$

thus in a ranked poset all rank rows are antichains. In a pioneering 1927 paper E. Sperner [22] showed that for $P = (2^{[n]}, \subseteq)$

The size of any antichain of P does not exceed the size of the largest rank row.

Because of this it has become customary to call a poset P with this property a *Sperner* poset.

In a remarkable 1980 paper R. Stanley [23] proved that the lattice $L[m, n]$ of Ferrers diagrams contained in the $m \times n$ rectangle, ordered by inclusion, is a Sperner poset. Stanley obtained this result using Algebraic Geometry. Later it was shown by R. Proctor [18] that this result may be obtained as a beautiful application of $\mathfrak{sl}(2)$ theory. Also of interest is Proctor's expository article on the Sperner property for a number of combinatorially interesting cases that appeared in 1982 [19]. Our goal in this section is to derive the Spernerity of $L[m, n]$ by from our development of $\mathfrak{sl}(2)$ theory.

But before we can proceed we need to make some preliminary observations and derive a few auxiliary facts.

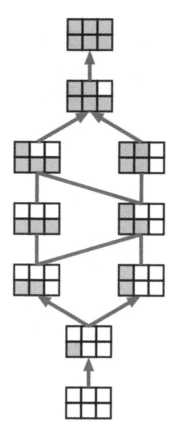

(3.3.8)

In Fig. (3.3.8) we have depicted the Hasse diagram of the lattice $L[2, 3]$. It is easily seen from this example that $L[m, n]$ is a ranked poset, with the rank of a diagram $\pi \in L[m, n]$ given by the function,

$$r(\pi) = \#\,\text{cells of } \pi .\tag{3.3.9}$$

It is easily seen that $L[2, 3]$ is also rank symmetric and rank unimodal. Our first task is to show that this is true in full generality for all $L[m, n]$.

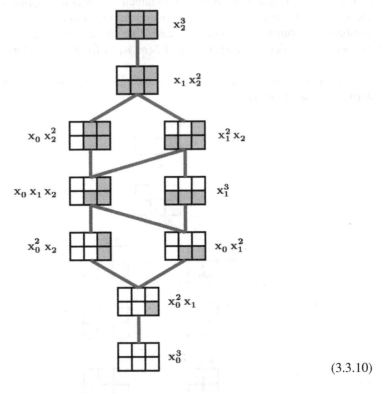

$$(3.3.10)$$

It will be convenient to slightly alter the way we depict $L[m, n]$ as we have done for $L[2, 3]$ in Fig. (3.3.10). Here we have depicted $L[2, 3]$ with its elements represented by monomials. This will play a crucial role in our development. To be consistent with monomial representation, we have depicted the Ferrers diagrams with right justified rows.

This given we begin by showing that the rank generating function of $L[m, n]$ is given by the q-binomial polynomial. More precisely we have

Proposition 3.10

$$F_{L[m,n]}(q) = \begin{bmatrix} m + n \\ n \end{bmatrix} = \frac{\prod_{i=1}^{m+n}(1 - q^i)}{\prod_{i=1}^{m}(1 - q^i) \prod_{i=1}^{n}(1 - q^i)}. \tag{3.3.11}$$

Proof For $\pi \in L[m, n]$ we set (as we did for $L[2, 3]$):

$$m[\pi] = x_0^{p_0} x_1^{p_1} \cdots x_m^{p_m} \tag{3.3.12}$$

where p_i gives the number of columns of π that are of length i. Now the generating function of all these monomials is given by the formal series

$$\sum_{p_0 \geq 0} \sum_{p_1 \geq 0} \cdots \sum_{p_m \geq 0} t^{p_0 + p_1 + \cdots + p_m} x_0^{p_0} x_1^{p_1} \cdots x_m^{p_m} \tag{3.3.13}$$

$$= \frac{1}{(1 - tx_0)(1 - tx_1) \cdots (1 - tx_m)}.$$

Since the number of columns of π is given by the sum $p_0 + p_1 + \cdots + p_m$ the coefficient of t^n in (3.3.13) yields the list of all the monomials that correspond to elements of $L[m, n]$. Now we see from the definition in (3.3.9) that we have $r(\pi) = \sum_{i=1}^{m} i \, p_i$ and (3.3.11) gives

$$m[\pi] \Big|_{x_i \to q^i} = x_0^{p_0} x_1^{p_1} \cdots x_m^{p_m} \Big|_{x_i \to q^i} = q^{r(\pi)}.$$

Making the replacements $x_i \to q^i$ on both sides of (3.3.13) immediately yields the beautiful identity

$$\sum_{n \geq 0} t^n F_{L[m,n]}(q) = \frac{1}{(1 - t)(1 - tq) \cdots (1 - tq^m)}. \tag{3.3.14}$$

The assertion in (3.3.11) is then an immediate consequence of the so called q-binomial identity. We shall include a proof of it here for sake of completeness. To this end recall that the q-analogue of differentiation is defined by setting for any formal series $f(t)$

$$D_q f(t) = \frac{f(t) - f(qt)}{t(1 - q)}. \tag{3.3.15}$$

Now an easy calculation gives that

$$D_q t^k = [k]_q t^{k-1} \tag{3.3.16}$$

and

$$D_q \frac{1}{(1-t)(1-tq)\cdots(1-tq^{k-1})} = \frac{[k]_q}{(1-t)(1-tq)\cdots(1-tq^k)}, \tag{3.3.17}$$

where, as customary we have set

$$[k]_q = \frac{1-q^k}{1-q} = 1 + q + q^2 + \cdots + q^{k-1}. \tag{3.3.18}$$

Using (3.3.17) recursively for $k = 1, 2, \ldots, m-1$ yields that

$$\frac{1}{[m]_q!} D_q^m \frac{1}{1-t} = \frac{1}{(1-t)(1-tq)\cdots(1-tq^m)}. \tag{3.3.19}$$

On the other hand from (3.3.16) we derive that we must also have

$$D_q^m \frac{1}{1-t} = \sum_{k\geq m} [k]_q [k-1]_q \cdots [k+1]_q \, t^{k-m}.$$

By changing the summation index to $n = m - k$ we can also rewrite this as

$$\frac{1}{[m]_q!} D_q^m \frac{1}{1-t} = \sum_{n\geq 0} \frac{[m+n]_q [m+n-1]_q \cdots [n+1]_q}{[m]_q!} \, t^n \tag{3.3.20}$$

$$= \sum_{n\geq 0} \begin{bmatrix} n+m \\ m \end{bmatrix}_q t^n .$$

Equating the righthand sides of (3.3.19) an (3.3.20) we finally obtain the so-called q-binomial identity

$$\frac{1}{(1-t)(1-tq)\cdots(1-tq^m)} = \sum_{n\geq 0} \begin{bmatrix} n+m \\ m \end{bmatrix}_q t^n . \tag{3.3.21}$$

Using this in (3.3.14) and equating coefficients of t^n gives (3.3.11) and completes our proof. $\qquad\square$

Note that from (3.3.11) we get

$$F_{L[2,3]}(q) = \begin{bmatrix} 2+3 \\ 2 \end{bmatrix} = \frac{(1-q^4)(1-q^5)}{(1-q)(1-q^2)} = 1 + q + 2q^2 + 2q^3 + 2q^4 + q^5 + q^6$$

which is in perfect agreement with what we infer from our drawing of $L[2, 3]$.

To apply $\mathfrak{sl}(2)$ theory to the study of the $L[m, n]$ posets we need to view the elements of $L[m, n]$ as bases of a vector space. More precisely we shall work with the vector space $V_{m,n}$ which is the linear span over $\mathbf{Q}$ of the set of monomials

$$\mathcal{B}_{m,n} = \left\{ m(\pi) = x_0^{p_0} x_1^{p_1} \cdots x_m^{p_m} \mid \pi \in L[m, n] \right\}.$$

To construct an $\mathfrak{sl}(2)$ action on this space we need a few preliminary observations.

We shall begin with the following result which is of independent interest.

Proposition 3.11 *Given an $m \times m$ matrix $A = \|a_{ij}\|_{i,j=0}^m$ let D_A denote the differential operator*

$$D_A = \sum_{i,j=0}^m x_i a_{ij} \partial_{x_j} . \tag{3.3.22}$$

It is easy to see that D_A acts as ordinary differentiation on polynomials in $x_0, x_1, \ldots, x_m$. That is for $P, Q \in \mathbf{Q}[x_0, x_1, \ldots, x_m]$ we have

$$D_A(PQ) = (D_A P)Q + P D_A Q .$$

But more importantly the map $A \to D_A$ preserves the bracket operation. That is

$$\left[D_A, D_B \right] = D_{[A,B]} . \tag{3.3.23}$$

Proof If we let $B = \|b_{ij}\|_{i,j=0}^m$, then for any polynomial P we have

$$D_A D_B P = \sum_{i_1, j_1=0}^m x_{i_1} a_{i_1, j_1} \partial_{x_{j_1}} \sum_{i_2, j_2=0}^m x_{i_2} b_{i-2, j_2} \partial_{x_{j_2}} P$$

$$= \sum_{i_1, j, i_2=0}^m x_{i_1} a_{i_1, j} b_{j, j_2} \partial_{x_j} x_j \partial_{x_{j_2}} P + \sum_{i_1, j_1 \neq i_2, j_2} x_{i_1} x_{i_2} a_{i_1, j_1} b_{i_2, j_2} \partial_{x_{j_1}} \partial_{x_{j_2}} P$$

$$= D_{AB} P + \sum_{i_1, j_1, \neq i_2, j_2} x_{i_1} x_{i_2} a_{i_1, j_1} b_{i_2, j_2} \partial_{x_{j_1}} \partial_{x_{j_2}} P .$$

Repeating this calculation with the roles of A and B interchanged and subtracting we see that the second terms on the right hand sides cancel and the first terms yield the right hand side of (3.3.23) as desired. $\square$

This proposition implies that if $\mathcal{G}$ is a Lie algebra of $(m+1) \times (m+1)$ matrices then the operators D_A with $A \in \mathcal{G}$ yield a representation of $\mathcal{G}$ on the space of polynomials in the variables $x_0, x_1, \ldots, x_m$. In particular we can obtain a representation of $\mathfrak{sl}(2)$ on the space of polynomials in x, y by means of the operators that correspond to the matrices

$$E = \begin{pmatrix} 0 & 1 \\ 0 & 0 \end{pmatrix}, \quad F = \begin{pmatrix} 0 & 0 \\ 1 & 0 \end{pmatrix}, \quad H = \begin{pmatrix} 1 & 0 \\ 0 & -1 \end{pmatrix}.$$

We see that in this case the definition in (3.3.22) with $m = 1$ $x_0 = x$ and $x_1 = y$ gives

$$D_E = x\partial_y, \qquad D_F P(x, y) = y\partial_x, \qquad D_H = (x\partial_x - y\partial_y).$$

Let us now restrict this action to homogeneous polynomials of degree m:w in x, y and express it in terms of the monomial basis

$$\left\{ x_j = x^j y^{m-j} \mid j = 0, 1, \ldots, m \right\}. \tag{3.3.24}$$

If we agree to set

$$x_{-1} = x_{m+1} = 0$$

a simple calculation yields that for $i = 0, 1, \ldots, m$ we have

$$D_E x_j = (m-j)x_{j+1}, \quad D_F x_j = j\,x_{j-1}, \quad D_H x_j = (2j - m)x_i. \tag{3.3.25}$$

We thus see that this construction brings us back to the matrices $\tilde{E}, \tilde{F}, \tilde{H}$ given in (3.2.56). We thus obtain another proof that these matrices satisfy the relations

$$[\tilde{E}, \tilde{F}] = \tilde{H}, \quad [\tilde{H}, \tilde{E}] = 2\tilde{E}, \quad [\tilde{H}, \tilde{F}] = -2\tilde{F},$$

It thus follows, from Proposition 3.11 that the differential operators

$$(a)\ D_{\tilde{E}} = \sum_{j=0}^{m-1} (m-j)x_{j+1}\partial_{x_j},$$

$$(b)\ D_{\tilde{F}} = \sum_{j=1}^{m} j\,x_{j-1}\partial_{x_j}, \tag{3.3.26}$$

$$(c)\ D_{\tilde{E}} = \sum_{j=0}^{m} (2j - m)x_j\partial_{x_j}$$

yield a representation of $\mathfrak{sl}(2)$ on the polynomials in $x_0, x_1, \ldots, x_m$.

Now it develops that we have the following remarkable identities:

Proposition 3.12 *The action of the operators $D_{\tilde{E}}, D_{\tilde{F}}, D_{\tilde{H}}$ on the monomial basis* $\{m(\pi) \mid \pi \in L[m.n]\}$ *may be expressed in the form*

$$(a) \quad D_{\tilde{E}}m(\pi) = \sum_{\pi^+} m(\pi^+) \chi(\pi \to \pi^+) w(\pi, \pi^+)$$

$$(b) \quad D_{\tilde{F}}m(\pi) = \sum_{\pi^-} m(\pi^-) \chi(\pi^- \to \pi) w(\pi, \pi^-) \qquad (3.3.27)$$

$$(c) \quad D_{\tilde{H}}m(\pi) = \big(2r(\pi) - mn\big) m(\pi)$$

where $w(\pi, \pi^+)$ and $w(\pi, \pi^-)$ denote suitable integer weights.

Proof Note that for $\pi \in L[m, n]$ we may write

$$D_{\tilde{E}}m(\pi) = \sum_{j=0}^{m-1}(m-j)x_{j+1}\partial_{x_j}\, x_0^{p_0}x_1^{p_1}\cdots x_m^{p_m}$$

$$= \sum_{j=0}^{m-1}(m-j)p_j\, \tfrac{x_{j+1}}{x_j}\, m(\pi) .$$

But when $p_j > 0$, the monomial $\tfrac{x_{j+1}}{x_j} m(\pi)$ is none other than the monomial of the partition obtained by removing from π a column of length j and adding a column of length $j + 1$. The resulting partition π^+ is easily seen to be one of the successors of π in the $L[m, n]$ poset. This proves (3.3.27) (a) with

$$w(\pi, \pi^+) = (m-j)p_j .$$

Similarly from (3.3.26) (b) we derive that

$$D_{\tilde{F}}m(\pi) = \sum_{j=1}^{m} j\, x_{j-1}\partial_{x_j}\, x_0^{p_0}x_1^{p_1}\cdots x_m^{p_m}$$

$$= \sum_{j=1}^{m} j\, p_j\, \tfrac{x_{j-1}}{x_j}\, m(\pi) .$$

Here again, when $p_j > 0$, the monomial $\tfrac{x_{j-1}}{x_j} m(\pi)$ is none other than the monomial of the partition obtained by removing from π a column of length j and adding a column of length $j - 1$. The resulting partition π^- is easily seen to be one of the predecessors of π in the $L[m, n]$ poset. This proves (3.3.27) (b).

In the same vein (3.3.26) (c) gives

$$D_{\tilde{H}} m(\pi) = \sum_{j=0}^{m} (2j - m) \, x_j \partial_{x_j} \, x_0^{P_0} x_1^{P_1} \cdots x_m^{P_m} \qquad (3.3.28)$$

$$= \sum_{j=0}^{m} (2j - m) \, p_j \, m(\pi) \, .$$

But when $\pi \in L[m, n]$ we have the relations

$$\sum_{j=0}^{m} jp_j = area(\pi), \quad \sum_{j=0}^{m} p_j = n \, .$$

Since, as we have noted that $area(\pi) = r(\pi)$ we see that (3.3.27) is simply another way of writing (3.3.27) (c). This completes our proof.   □

At this point it is good to shorten the symbol introduced in (3.3.2) and simply use $\mathcal{R}_s$ to denote the sth rank row of $L[m, n]$. In other words $\mathcal{R}_s$ consists of all the partitions $\pi \in L[m, n]$ of area s, We will use n_s to denote the cardinality of $\mathcal{R}_s$. It is also convenient to denote by π^c the partition corresponding to the complement of π in the $m \times n$ rectangle.

We are finally in a position to apply $\mathfrak{sl}(2)$ theory to the study of $L[m, n]$. We begin with the following result.

Theorem 3.13 *The $L[m, n]$ poset is rank symmetric and rank unimodal. In particular the coefficients of the q-binomial polynomials form a unimodal sequence.*

Proof We see from (3.3.27) (c) that all the monomials $m(\pi)$ are eigenvectors of $D_{\tilde{H}}$. Since these monomials are a basis of $V_{m,n}$, it follows from (3.3.27) that the eigenspace V_{-k} of $D_{\tilde{H}}$ corresponding the eigenvalue $-k \leq 0$ is given by the linear span

$$V_{-k} = \mathcal{L}[m(\pi) \mid \pi \in \mathcal{R}_{(mn-k)/2}] \, .$$

Thus we immediately derive from Theorem 3.9 that

$$n_{s-1} \leq n_s \qquad \text{(for } s \leq mn/2)$$

and

$$n_{s+1} \leq n_s \qquad \text{(for } s \geq mn/2) \, .$$

This proves rank unimodality. Rank symmetry immediately follows from the simple fact that the map $\pi \to \pi^c$ is an involution of $L[m, n]$ which maps a rank row onto a rank row of complementary rank. Nevertheless it is interesting to note that Theorem 3.8 shows that also rank symmetry follows from $\mathfrak{sl}(2)$ theory. □

Theorem 3.14 $L[m, n]$ *is a Sperner poset.*

Proof Note first that if a poset P can be decomposed into the disjoint union of d chains then its largest antichain can have no more that d elements. This is clear since the intersection of an antichain with any chain can contain at most one element. Thus one of the standard ways of proving Spernerity is to produce a chain cover whose cardinality is equal to that of the largest rank row. It develops that $\mathfrak{sl}(2)$ theory provides us precisely with the tools we need to construct such a chain cover for $L[m, n]$. To proceed we need an auxiliary definition.

For a given $0 \leq s < k_0 = \lceil \frac{mn}{2} \rceil$, a map

$$\phi_s : \mathcal{R}_s \longrightarrow \mathcal{R}_{s+1}$$

will be called *admissible* if and only if

$$
\begin{aligned}
&(i) \quad \phi_s \text{ is injective,} \\
&(ii) \quad \text{for all } \pi \in \mathcal{R}_s \text{ we have } \pi \to \phi_s(\pi) .
\end{aligned}
\tag{3.3.29}
$$

Now note that if we are in possession of a sequence of admissible maps ϕ_0, $\phi_1, \ldots, \phi_s, \ldots, \phi_{k_0}$, then the construction of the desired chain cover is immediate. We simply erase from the Hasse diagram of $L[m, n]$ all containment arrows except those corresponding to the sets of pairs

$$\mathcal{A} = \bigcup_{0 \leq s \leq k_0} \left\{ \left(\pi, \phi_s(\pi)\right) ; \ \pi \in \mathcal{R}_s \right\} \tag{3.3.30}$$

$$\mathcal{B} = \bigcup_{0 \leq s < k_0} \left\{ \left(\pi^c, \phi_s(\pi)^c\right) ; \ \pi \in \mathcal{R}_s \right\} .$$

It is easily seen that the remaining arrows depict our chain cover. In Fig. (3.3.31) we illustrate such a construction for $L[3, 3]$. There the darker arrows come from the pairs in $\mathcal{B}$ of (3.3.30). It should be clear that the number of chains produced by this construction equals the number of elements in $\mathcal{R}_{k_0}$.

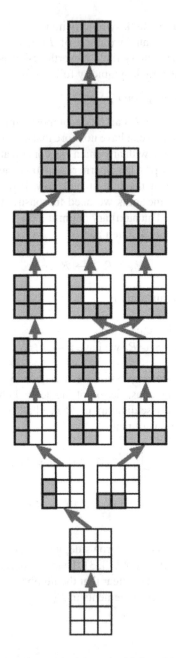

$$(3.3.31)$$

This given, our proof is completed by combining Theorem 3.9 with Proposition 3.12. In fact, it follows from Theorem 3.9 that, for each $s \le k_0$, the $n_s \times n_{s+1}$ matrix giving the action of $D_{\tilde{E}}$ on the monomials corresponding to the elements of rank s, must necessarily have rank equal to n_s. This implies that this matrix must contain an $n_s \times n_s$ nonvanishing subdeterminant. In turn this subdeterminant will have at least one nonvanishing term. But the nature of the matrix of $D_{\tilde{E}}$ as given by (3.3.27) (a) yields that any nonvanishing determinantal term yields a map of $\mathcal{R}_s$ into $\mathcal{R}_{s+1}$ which satisfies properties (i) and (ii) of (3.3.29). Choosing one of these injections for each rank $s \leq k_0$ and calling it ϕ_s yields the sequence of admissible maps which yields the chain cover that proves Spernerity.   □

Remark 3.3 A problem that has remained open since the early 80's is to prove the existence of a minimal chain cover for $L[m, n]$ consisting entirely of *saturated symmetric chains*. These are saturated chains which join a element of rank s to an element of rank $mn - s$. The decomposition of $V_{m,n}$ as a direct sum of strings resulting from the $\mathfrak{sl}(2)$ action may be viewed as an *algebraic level* form of saturated symmetric chain cover. Unfortunately, this argument falls short of providing such a cover at the *combinatorial level*, and it is easily seen from the previous figure what could cause lack of symmetry. Nevertheless the single replacement of the arrow $(2) \rightarrow (3)$ by the arrow $(2) \rightarrow (2, 1)$ yields a symmetric chain cover for $L[3, 3]$.

Lecture 4
Heaps, Continued Fractions, and Orthogonal Polynomials

4.1 Introduction

These notes are an exposition of work of Françon–Viennot [9], Flajolet [8] and Viennot [25, 26] dealing with combinatorial aspects of the theory of continued fractions. We have added here some of the missing details, some of the proofs that were not given in the original works, and hopefully have corrected some of the errors without creating new ones. We also give a brief review of the classical connection between orthogonal polynomials and continued fractions.

4.2 Heaps of Monomers and Dimers

Before making definitions it is best to start with an example. In Fig. (4.2.1) we have an instance of a *monomer-dimer heap*.

$$(4.2.1)$$

Let us imagine that the vertical lines represent needles and that there are two basic pieces:

A *monomer*: which is just a ring, and
A *dimer*: which consists of two rings joined by a metal bar.

© The Editor(s) (if applicable) and The Author(s), under exclusive license to Springer Nature Switzerland AG 2020
A. M. Garsia, Ö. Eğecioğlu, *Lectures in Algebraic Combinatorics*,
Lecture Notes in Mathematics 2277, https://doi.org/10.1007/978-3-030-58373-6_4

To put together a heap we simply pick a bunch of monomers and dimers and stack them on top of each other by threading them into the needles as indicated in Fig. (4.2.1) above. As it is indicated there the needles are perpendicular to the ground line at the points of coordinates $0, 1, 2, 3 \dots$. Heaps of monomers and dimers will be represented here by words in the alphabet

$$\mathcal{A} = \{m_0, m_1, m_2, \dots, d_1, d_2, d_3, \dots\}$$

by replacing each monomer of ground coordinate i by the letter m_i and each dimer projecting onto the interval $[i - 1, i]$ by the letter d_i. The corresponding word is obtained by processing in this manner the successive pieces of the heap from left to right within a row, starting from the bottom row and proceeding upwards. For instance, this procedure applied to the heap of Fig. (4.2.1) yields the word

$$w = d_2 m_6 m_7 d_1 d_3 m_4 d_6 m_0 d_2 d_4 m_5 m_2 \ .$$

Conversely, given any word $w \in \mathcal{A}^*$ we can construct a heap by reversing the process above. That is we read the letters of w from left to right and replace each m_i by a monomer of ground coordinate i and each d_i by a dimer spanning $[i - 1, i]$. Of course we must also thread the corresponding monomers and dimers down the needles in the precise succession they are encountered as we read w. The final configuration is obtained by letting the pieces settle as far down as they can. This procedure applied to the word $w_1 = m_0 d_2 m_2 d_1 m_1 d_2 m_3 m_3$ produces the heap in Fig. (4.2.2).

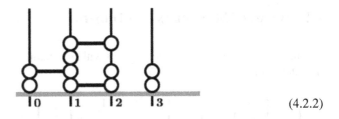

$$(4.2.2)$$

Now it is easy to see that the word $w_2 = m_3 d_2 m_0 d_1 m_3 m_1 m_2 d_2$ produces the same heap. At the same time the word which corresponds to this heap by the construction given above shoud be $w = m_0 d_2 m_3 d_1 m_2 m_3 m_1 d_2$. Mathematically speaking a heap should be represented by an equivalence class of $\mathcal{A}$-words. Two words being equivalent if and only if they yield the same heap. Thus our procedure of constructing the word corresponding to a heap is just one of the ways of selecting a representative from each equivalence class of words.

Imagine now that the needles of Fig. (4.2.1) are set into a top and bottom bar as in an abacus and we pull down the monomer m_2. This will result in the configuration in Fig. (4.2.3).

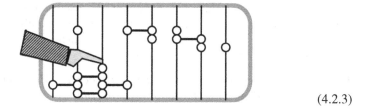

(4.2.3)

We can see that there are heaps that are brought down by pulling on a single piece. Such is also the case for the heap obtained by removing the monomers m_0 and m_7 from the heap in (4.2.3) and adding on top the dimer d_4. Such a heap will be called a *pyramid* and the top piece the *summit* of the pyramid.

There is a very simple way of transforming Motzkin paths into heaps which, as we shall see, has remarkable mathematical consequences. This transformation is obtained by replacing each EAST step in the path by a monomer and each NORTHEAST step by a dimer. We only need one example here to get across what we have in mind. For instance, carrying out these replacements (from left to right) on the path on the left in Fig. (4.2.4) yields the configuration at the bottom.

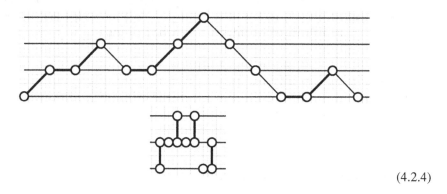

(4.2.4)

The latter is then rotated $90°$ clockwise so that the pieces settle down to the ground. This results in the heap given in Fig. (4.2.5):

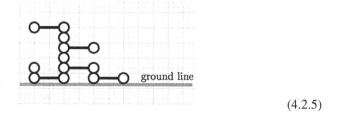

ground line

(4.2.5)

The transformation is even simpler at the word level. In fact, the a, b, c-word corresponding to the Motzkin path is $w = a_1c_1a_2b_1c_1a_2a_3b_2b_1b_0c_0b_1b_0$ while the word of the heap is $w' = d_1 m_1 d_2 m_1 d_2 d_3 m_0 d_1$. We see that to go from the word w of the path to the word w' of the corresponding heap, we simply read the letters of w from right to left, replace each a by a d, each c by an m and remove all b's. From our example we can easily extract the following

Theorem 4.1 *Our construction yields a bijection between Motzkin paths and heaps of monomers and dimers with the following properties:*

1. *The image heap is always a pyramid with summit a monomer m_0 or a dimer d_1.*
2. *Paths whose maximum height does not exceed n correspond to pyramids whose projection is in the interval $[0, n]$.*
3. *Dyck paths are sent into pyramids of dimers with summit d_1.*
4. *If the image pyramid has d dimers and m monomers then the corresponding path has $2d + m$ steps.*

4.3 The Cartier–Foata Languages

There is a purely algebraic way to define the equivalence classes of words yielding the same monomer-dimer heap. Let $\mathcal{A} = \{a_1, a_2, \ldots, a_n\}$ be an alphabet and let $M = \|m_{ij}\|_{i,j=1}^n$ be a matrix with entries $0, 1$ with the properties

$$m_{ii} = 1 \quad \text{and} \quad m_{ij} = m_{ji} \ .$$

Let us say that two letters a_i and a_j *commute* if and only if $m_{ij} = 0$. Given two words $w_1, w_2 \in \mathcal{A}^*$ let us write

$$w_1 \longleftrightarrow_M w_2$$

if and only if we have the two factorizations

$$w_1 = w' \, a_i a_j \, w'' \ , \qquad w_2 = w' \, a_j a_i \, w''$$

with a_i and a_j a pair of commuting letters. Finally, let us write $w_1 \approx_M w_2$ if and only if we have a sequence of words $\sigma^{(i)} \in \mathcal{A}^*$, $(i = 1, 2, \ldots, n)$ such that

$$w_1 = \sigma^{(1)} \longleftrightarrow \sigma^{(2)} \longleftrightarrow \cdots \longleftrightarrow \sigma^{(n-1)} \longleftrightarrow \sigma^{(n)} = w_2 \ .$$

In words, $w_1 \approx_M w_2$, if and only if w_2 can be obtained from w_1 by a sequence of transpositions of pairs of adjacent commuting letters. The quotient of $\mathcal{A}^*$ by the equivalence $\approx_M$ will be referred to as the *Cartier–Foata language* corresponding to the pair $(\mathcal{A}, M)$. Clearly the equivalence $\approx_M$ is compatible with concatenation of words. That is if $w_1 \approx_M w_1'$ and $w_2 \approx_M w_2'$ then $w_1 w_2 \approx_M w_1' w_2'$. Thus the concatenation product gives $\mathcal{A}^* / \approx_M$ the structure of a monoid.

Given a pair $(\mathcal{A}, M)$ it will be convenient to denote by $\mathcal{L}(\mathcal{A}, M)$ the formal sum of a suitably selected set of representatives for the equivalence classes of $\mathcal{A}^*$ modulo $\approx_M$. When dealing with a Cartier–Foata language, all identities between formal sums of words of $\mathcal{A}^*$ where the equal sign is subscripted by M should be understood to hold modulo the equivalence $\approx_M$. Clearly, to define a Cartier–Foata language we need only specify which are the commuting pairs. Indeed, this is the only information yielded by the matrix M. This given, we can easily see that if we take

$$\mathcal{A} = \{m_0, m_1, m_2, \ldots, d_1, d_2, d_3, \ldots\} \tag{4.3.1}$$

and declare that the commuting pairs are

$$\begin{aligned}
&(m_i, m_j) \text{ for } i \neq j, \\
&(m_i, d_j) \text{ for } \{i\} \cap \{j-1, j\} = \emptyset, \\
&(d_i, d_j) \text{ for } \{i-1, i\} \cap \{j-1, j\} = \emptyset
\end{aligned} \tag{4.3.2}$$

then the corresponding Cartier–Foata language may be identified with the collection of heaps of monomers and dimers.

As we shall see we can associate to each Cartier–Foata language $\mathcal{L}(\mathcal{A}, M)$ a collection of heap-like objects which may be represented by the words of $\mathcal{L}(\mathcal{A}, M)$. Nevertheless, these languages are of independent interest and it will be good to start by establishing some of the basic identities concerning them. To this end we need to make some definitions.

For a given word $w \in \mathcal{L}(\mathcal{A}, M)$ set

$$I(w) = \{a \in \mathcal{A} \mid w \approx_M a w'\}. \tag{4.3.3}$$

In words, $I(w)$ represents the set of letters that may be brought to the front of w by a sequence of transpositions of pairs of adjacent commuting letters. For this reason we shall refer to $I(w)$ as the *initial part* of w. Similarly we define as the *end part* of w the set

$$E(w) = \{b \in \mathcal{A} \mid w \approx_M w' b\}. \tag{4.3.4}$$

In the monomer-dimer case (that is when the language is given by (4.3.1) and (4.3.3)) a word w corresponds to a pyramid if and only if $E(w)$ has only one element. It will be convenient to extend our heap terminology to the general Cartier–Foata language and call *pyramid* any word whose $E(w)$ is a singleton. Note that if $E(w)$ is not a singleton the any two letters occurring in $E(w)$ must form a commuting pair. Of course the same holds true for $I(w)$. Subsets of $\mathcal{A}$ consisting of pairwise commuting letters shall be referred to as *free* subsets. their collection will be denoted by $\mathcal{F}(\mathcal{A}, M)$. Given a free set F we shall also denote by F the word obtained by multiplying the letters of F in alphabetic order. Of course since all pairs of letters of F commute, whatever order we use produces the same element of the language so this abuse of notation cannot lead to ambiguities. With these conventions, the basic result of the theory of Cartier–Foata languages may be stated as follows as follows:

Theorem 4.2 (Cartier–Foata [5]) *Modulo the equivalence $\approx_M$, the generating series of the language may be computed from the following identity*

$$\mathcal{L}(\mathcal{A}, M) \ =_M \ \frac{1}{\sum_{F \in \mathcal{F}(\mathcal{A}, M)} (-1)^{|F|} \, F} \tag{4.3.5}$$

where $|F|$ denotes here the cardinality of F.

Proof We only need to show that the product of the left hand side by the denominator of the right hand side evaluates to the empty word. To this end we write this product in the form

$$\sum_{w \in \mathcal{A}^*} w \sum_{F \in \mathcal{F}(\mathcal{A}, M)} \sum_{v \in \mathcal{L}(\mathcal{A}, M)} (-1)^{|F|} \, \chi(w \approx_M F \, v) \ . \tag{4.3.6}$$

Now, since F is a free set, the condition that $w = Fv$ is equivalent to the statement that $F \subseteq I(w)$. Thus we see that the inner sum may be simply rewritten as

$$\sum_{F \subseteq I(w)} (-1)^{|F|}$$

and this clearly evaluates to zero unless of course w itself reduces to the empty word. This establishes our identity.   □

This result has a number of interesting corollaries and applications. For instance let $c_n = c_n(\mathcal{A}, M)$ denote the number of words of length n in $\mathcal{L}(\mathcal{A}, M)$ and let $F_{\mathcal{A}, M}(t)$ denote the generating functions of the sequence $\{c_n\}$ then from (4.3.6) we immediately deduce the following

Proposition 4.1

$$F_{\mathcal{A}, M}(t) = \frac{1}{\sum_{F \in \mathcal{F}(\mathcal{A}, M)} (-t)^{|F|}} \ . \tag{4.3.7}$$

In particular, for heaps of monomer and dimers on the interval $[0, 2]$ there are twelve free sets. Namely, the empty set, the singletons

$$\{m_0\}, \{m_1\}, \{m_2\}, \{d_1\}, \{d_2\}$$

the pairs

$$\{m_0, m_1\}, \{m_0, m_2\}, \{m_1, m_2\}, \{m_0, d_2\}, \{d_1, m_2\}$$

and the triplet $\{m_0, m_1, m_2\}$. This gives, that in this case

$$F_{\mathcal{A}, M}(t) = \frac{1}{1 - 5t + 5t^2 - t^3} . \qquad (4.3.8)$$

A curious feature of the identity in (4.3.7) is that it provides a tool for proving that certain rational functions have Taylor series with nonnegative integral coefficients, a fact that, even in such a simple instance as (4.3.8), is not easily established.

Theorem 4.2 has a very useful refinement. Given a subalphabet $\mathcal{B} \subseteq \mathcal{A}$ we shall let $\mathcal{L}(\mathcal{B}, M)$ denote the sublanguage consisting of the words of $\mathcal{L}(\mathcal{A}, M)$ all of whose letters are in $\mathcal{B}$. At the same time we shall denote by $\mathcal{L}^{(\mathcal{B})}(\mathcal{A}, M)$ the subset of $\mathcal{L}(\mathcal{A}, M)$ consisting of all the words of $\mathcal{L}(\mathcal{A}, M)$ whose end part lies entirely in $\mathcal{B}$. This given, we have the following result.

Theorem 4.3 (De Sainte–Catherine [26])

$$\mathcal{L}^{(\mathcal{B})}(\mathcal{A}, M) =_M \frac{\sum_{F \in \mathcal{F}(\mathcal{A} - \mathcal{B}, M)} (-1)^{|F|} F}{\sum_{F \in \mathcal{F}(\mathcal{A}, M)} (-1)^{|F|} F} . \qquad (4.3.9)$$

Proof Using Theorem 4.2 (twice) this identity may be rewritten in the form

$$\mathcal{L}^{(\mathcal{B})}(\mathcal{A}, M) \times \mathcal{L}(\mathcal{A} - \mathcal{B}, M) =_M \mathcal{L}(\mathcal{A}, M) . \qquad (4.3.10)$$

In turn this is equivalent to stating that every word of $\mathcal{L}(\mathcal{A}, M)$ may be uniquely factorized (modulo $\approx_M$) into the concatenation product of a word of $\mathcal{L}^{(\mathcal{B})}(\mathcal{A}, M)$ by a word of $\mathcal{L}(\mathcal{A} - \mathcal{B}, M)$. To see that this is indeed so, imagine that the letters of a word $w \in \mathcal{L}(\mathcal{A}, M)$ are strung as beads into a long string. Imagine further that pairs of letters that commute can *pass through each other*. This given the desired factorization can simply be obtained by pulling to the left all the *beads* of w that are in $\mathcal{B}$. This splits w into two words w_1 and w_2. Where w_1 consists of all the beads that had to move as a result of the pull and w_2 consists of all the beads to the right of the beads in $\mathcal{B}$ plus those that passed through them. We can also construct this factorization by a more standard mathematical argument. Let $w = a_1 a_2 \cdots a_i a_{i+1} \cdots a_n$ where a_i is the rightmost letter of w that is in $\mathcal{B}$. Then let $u = a_1 a_2 \cdots a_i$ and $v = a_{i+1} \cdots a_n$. We can now factorize the end part of u (modulo $\approx_M$) in the form $E(u) = \beta\rho$ where β consists of all the letters of $E(u)$ that are in $\mathcal{B}$ and ρ has all the remaining letters. Now, if $u \approx_M u' E(u)$ we can set $w_1 = u'\beta$ and $w_2 = \rho v$ and obtain that $w \approx_M w_1 w_2$ with $E(w_1) \in \mathcal{B}^*$ and $w_2 \in (\mathcal{A} - \mathcal{B})^*$ as desired. The uniqueness is clear, since if $w \approx_M w_1' w_2'$ with $E(w_1') \in \mathcal{B}^*$ and $w_2' \in (\mathcal{A} - \mathcal{B})^*$ then we must necessarily have that $E(w_1') = \beta$, for (in passing from w to $w_1' w_2'$) no letter of u' can be moved past a letter of β. But this yields that $w_1' = u'\beta$ and $w_2' = \rho v$.   □

It will be convenient to denote by $\mathcal{P}(\mathcal{A}, M)$ the collection of pyramids of the Cartier–Foata language $\mathcal{L}(\mathcal{A}, M)$. It develops that pyramids are the building blocks of the language. The precise meaning of this assertion will emerge in the proof of the following result. To begin with let us use the symbol $\equiv_c$ to denote equality of two formal sums of words, under the condition that the letters are allowed to unconditionally commute. In other words ordinary equality in the formal power series sense should hold if each word on either side is replaced by a monomial in which the letters appearing in the word are rearranged in increasing order. This given, we have the following remarkable identity.

Theorem 4.4 (Viennot [26])

$$\sum_{\pi \in \mathcal{P}(\mathcal{A}, M)} \frac{\pi}{|\pi|} \equiv_c \log\big(\mathcal{L}(\mathcal{A}, M) \big) \tag{4.3.11}$$

where $|\pi|$ denotes the length of π.

Proof This turns out to be a beautiful application of the theory of exponential structures. Given a word $w = a_1 a_2 \cdots a_n \in \mathcal{L}(\mathcal{A}, M)$, and a permutation $\sigma = \sigma_1 \sigma_2 \cdots \sigma_n$, we set

$$P(w, \sigma) = \begin{pmatrix} a_1 \\ \sigma_1 \end{pmatrix} \begin{pmatrix} a_2 \\ \sigma_2 \end{pmatrix} \cdots \begin{pmatrix} a_n \\ \sigma_n \end{pmatrix} \tag{4.3.12}$$

and consider it as a word in the alphabet $P(\mathcal{A})$ whose letters are the pairs $\binom{a}{k}$ with $a \in \mathcal{A}$ and $k = 1, 2, \ldots$. We then define a new Cartier–Foata language $\mathcal{L}(P(\mathcal{A}), M)$ by letting two letters $\binom{a_i}{h}$ and $\binom{a_j}{k}$ of $P(\mathcal{A})$ commute if and only if a_i and a_j themselves commute (that is $m_{ij} = 0$). Using the *bead* imagery of the previous proof, we can factorize the word $P(w, \sigma)$ into a product of pyramids of the language $\mathcal{L}(P(\mathcal{A}), M)$. We simply thread the letters of $P(w, \sigma)$ on an infinite string, then pull to the left the letter $\binom{a_{j_1}}{1}$ of $P(w, \sigma)$. This will cause a number of letters of $P(w, \sigma)$ to move to the left as well. Again, a number of letters of $P(w, \sigma)$ that were originally to the left of $\binom{a_{j_1}}{1}$ will stay put. Those are the remaining letters of the end part of the left factor of $P(w, \sigma)$ terminating with $\binom{a_{j_1}}{1}$. By our convention those beads allow $\binom{a_{j_1}}{1}$ to pass through them. The letters that move with $\binom{a_{j_1}}{1}$ form a word that we would like to denote by $P(w_1, E_1)$. To do this we only need to extend our notation by allowing σ in (4.3.12) to be any sequence of natural numbers of the same length as w. Next, we set $i_1 = 1$ and let i_2 equal to the smallest integer in $\{1, 2, \ldots, n\}$ that is not in E_1. Now we locate the bead $\binom{a_{j_2}}{i_2}$ of $P(w, \sigma)$ and pull to the left on it moving with it every bead that cant let $\binom{a_{j_2}}{i_2}$ go through. This produces another factor $P(w_2, E_2)$. We can then let i_3 be the smallest integer in $\{1, 2, \ldots, n\}$ that is not in $E_1 + E_2$. Then drag to the left the letter $\binom{a_{j_3}}{i_3}$ of $P(w, \sigma)$. In this manner we successively construct the factorization

$$P(w, \sigma) \approx_M P(w_1, E_1)P(w_2, E_2)P(w_3, E_3) \cdots P(w_k, E_k) . \tag{4.3.13}$$

Our procedure guarantees the following three properties:

(1) Each factor $P(w_i, E_i)$ is a pyramid of $\mathcal{L}(\mathcal{A}, M)$,
(2) $i_1 = \min E_1 < i_2 = \min E_2 < \cdots < i_k = \min E_k$,
(3) $\{1, 2, \ldots, n\} = E_1 + E_2 + \cdots + E_k$, that is $(E_1, E_2, \ldots, E_k)$ is a partition of the set $\{1, 2, \ldots, n\}$.

Let now $\Pi_k = \Pi_k(\mathcal{A}, M)$ denote the collection of words $P(\pi, \tau)$ where

 (i) π is a pyramid of $\mathcal{L}(\mathcal{A}, M)$ of length k,
(ii) τ is a permutation in S_k with $\tau_k = 1$.

This given we see that the factorization in (4.3.13) exhibits the collection

$$\left\{ P(w, \sigma) \mid w \in \mathcal{L}(\mathcal{A}, M), \ \sigma \in S_n \right\}$$

as the exponential class constructed from the families Π_k $(k = 1, 2, \ldots)$. From the theory of exponential classes we then immediately derive that

$$\sum_{\substack{w \in \mathcal{L}(\mathcal{A}, M) \\ \sigma \in S_n}} \frac{P(w, \sigma)}{|w|!} \equiv_c \exp\left(\sum_{k \geq 1} \sum_{P(\pi, \tau) \in \Pi_k} \frac{P(\pi, \tau)}{k!} \right) . \tag{4.3.14}$$

To derive the identity in (4.3.12) we simply observe that the projection $P(w, \sigma) \to w$ removes the denominator $|w|!$ from (4.3.14), while the projection $P(\pi, \tau) \to \pi$, because of the restriction $\tau_k = 1$, only succeeds in reducing the denominator $k!$ to just k itself. Making these replacements in (4.3.14) yields

$$\sum_{w \in \mathcal{L}(\mathcal{A}, M)} w \equiv_c \exp\left(\sum_{k \geq 1} \sum_{\substack{\pi \in \mathcal{P}(\mathcal{A}, M) \\ |\pi| = k}} \frac{\pi}{k} \right) ,$$

and (4.3.11) follows by equating the logarithms of both sides. $\square$

4.4   Orthogonal Polynomials and Continued Fractions

Let $d\mu$ be a unit measure on $(-\infty, +\infty)$, and set

$$\int_{-\infty}^{+\infty} x^n \, d\mu = \mu_n \qquad (\mu_0 = 1) . \qquad (4.4.1)$$

The $\{\mu_n\}$'s are usually referred to as the *moments* of $d\mu$. They are well-defined for instance in the case that $d\mu$ is supported by some finite interval $[a, b]$, that is when $\mu(x) = 0$ for $x \leq a$ and $\mu(x) = \mu(b)$ for $x \geq b$). This permits us to define a scalar product on the space of polynomials in x by setting for any two polynomials $P(x), Q(x)$

$$\langle\, P\,,\; Q\,\rangle_\mu = \int_{-\infty}^{+\infty} P(x)Q(x) \, d\mu \; . \qquad (4.4.2)$$

It is easily seen that to compute this scalar product we only need to know the sequence of moments. Our object of study here is the sequence of polynomials $\{Q_n(x)\}_{n\geq 0}$ satisfying the following two conditions

$$(1) \;\; Q_n(x) = x^n + \sum_{k=0}^{n-1} a_{n,k}\, x^k$$

$$(2) \;\; \langle\, Q_n\,,\; Q_m\,\rangle_\mu = 0 \;\; \text{if} \;\; n \neq m \; . \qquad (4.4.3)$$

It develops that these polynomials have explicit expressions as ratios of determinants involving the μ_n's. More precisely we have

$$Q_n(x) = \frac{1}{d_{n-1}} \det \begin{pmatrix} \mu_0 & \mu_1 & \mu_2 & \cdots & \mu_n \\ \mu_1 & \mu_2 & \mu_3 & \cdots & \mu_{n+1} \\ \cdots & \cdots & \cdots & \cdots & \cdots \\ \mu_{n-1} & \mu_n & \mu_{n+1} & \cdots & \mu_{2n-1} \\ 1 & x & x^2 & \cdots & x^n \end{pmatrix} \qquad (4.4.4)$$

where

$$d_n = \det \begin{pmatrix} \mu_0 & \mu_1 & \mu_2 & \cdots & \mu_n \\ \mu_1 & \mu_2 & \mu_3 & \cdots & \mu_{n+1} \\ \cdots & \cdots & \cdots & \cdots & \cdots \\ \mu_{n-1} & \mu_n & \mu_{n+1} & \cdots & \mu_{2n-1} \\ \mu_n & \mu_{n+1} & \mu_{n+2} & \cdots & \mu_{2n} \end{pmatrix} . \qquad (4.4.5)$$

To prove this, it is sufficient to note that (4.4.4) gives

$$\langle Q_n, x^i \rangle = \frac{1}{d_{n-1}} \det \begin{pmatrix} \mu_0 & \mu_1 & \mu_2 & \cdots & \mu_n \\ \mu_1 & \mu_2 & \mu_3 & \cdots & \mu_{n+1} \\ \cdots & \cdots & \cdots & \cdots & \cdots \\ \mu_{n-1} & \mu_n & \mu_{n+1} & \cdots & \mu_{2n-1} \\ \mu_i & \mu_{i+1} & \mu_{i+2} & \cdots & \mu_{i+n} \end{pmatrix}$$

$$= \begin{cases} 0 & \text{if } i < n \\ \frac{d_n}{d_{n-1}} & \text{if } i = n . \end{cases} \tag{4.4.6}$$

In particular for any polynomial $P(x)$ of degree d we have the expansion

$$P(x) = \sum_{i \geq 0} \frac{\langle P, Q_i \rangle_\mu}{\langle Q_i, Q_i \rangle_\mu} Q_i(x) . \tag{4.4.7}$$

We should note that, because of (4.4.3) (1) and (2), the last nonvanishing summand on the right hand side of (4.4.7), is given by $i = \text{degree}(P)$. Note further that we must have

$$\langle x Q_n, x^i \rangle_\mu = \langle Q_n, x^{i+1} \rangle_\mu = 0 \qquad \text{(for } i = 0, 1, \ldots, n-2) ,$$

and this implies that must also have

$$\langle x Q_n, Q_i \rangle_\mu = 0$$

for all $i \leq n - 2$. Thus, when $P(x) = x Q_n(x)$ the expansion in (4.4.6) reduces to

$$x Q_n(x) = \frac{\langle x Q_n, Q_{n-1} \rangle_\mu}{\langle Q_{n-1}, Q_{n-1} \rangle_\mu} Q_{n-1}(x) +$$

$$+ \frac{\langle x Q_n, Q_n \rangle_\mu}{\langle Q_n, Q_n \rangle_\mu} Q_n(x) +$$

$$+ \frac{\langle x Q_n, Q_{n+1} \rangle_\mu}{\langle Q_{n+1}, Q_{n+1} \rangle_\mu} Q_{n+1}(x) . \tag{4.4.8}$$

Now, from (4.4.2) (1) and (4.4.6) we get that

$$\langle x Q_n, Q_{n+1} \rangle_\mu = \langle x^{n+1}, Q_{n+1} \rangle_\mu = \langle Q_{n+1}, Q_{n+1} \rangle_\mu . \tag{4.4.9}$$

Thus, setting

$$\lambda_n = \frac{\int_{-\infty}^{+\infty} x Q_{n-1}(x) Q_n(x) \, d\mu}{\int_{-\infty}^{+\infty} Q_{n-1}^2(x) \, d\mu} , \qquad c_n = \frac{\int_{-\infty}^{+\infty} x Q_n^2(x) \, d\mu}{\int_{-\infty}^{+\infty} Q_n^2(x) \, d\mu} , \tag{4.4.10}$$

we can rewrite (4.4.9) in the form

$$x Q_n(x) = \lambda_n \, Q_{n-1}(x) + c_n Q_n(x) + Q_{n+1}(x) \;,$$

or better yet

$$Q_{n+1} = (x - c_n) Q_n - \lambda_n \, Q_{n-1} \;. \tag{4.4.11}$$

It is customary in the theory of orthogonal polynomials to refer to this relation as the *three-term recursion* for the sequence $\{Q_n\}_{n \geq 0}$. Infact, starting from the trivial initial conditions

$$Q_{-1}(x) \equiv 0 \;, \qquad Q_0(x) \equiv 1 \tag{4.4.12}$$

we can recursively express each Q_n as a polynomial in x and the parameters $c_0, c_2, \ldots, c_{n-1}; \lambda_1, \lambda_2, \ldots, \lambda_{n-1}$. It develops that for most classical sequences of polynomials the recursion in (4.4.11) was known and could be easily derived before any measure $d\mu$ or sequence of moments was associated with them. Now, there is a Theorem, usually attributed to Favard [7], which asserts that a three term recursion is actually the necessary and sufficient condition for a sequence of polynomials $\{Q_n\}_{n \geq 0}$, satisfying (1) of (4.4.3), to be *orthogonal* with respect to *a sequence of moments*.

To be more precise, we should note that, given an arbitrary sequence $\{\mu_n\}_{n \geq 0}$, we can define a linear functional L_μ acting on polynomials in x by setting

$$L_\mu(x^n) = \mu_n \qquad \text{(for } n = 0, 1, 2, \ldots) \;. \tag{4.4.13}$$

We can then define a *scalar product* by setting

$$\langle P, Q \rangle_\mu = L_\mu(PQ) \;. \tag{4.4.14}$$

This given, Favard's Theorem says that if the Q_n's satisfy the recurrence (4.4.11) and the initial conditions (4.4.12) then they must satisfy the relations in (4.4.3) with a scalar product $\langle P, Q \rangle_\mu$ given by (4.4.14), (4.4.13) and a suitable sequence of μ_n's. Given the μ_n's, the existence of a *measure $d\mu$* satisfying (4.4.1) is usually referred to as the *moment problem* and it may be of considerable analytical difficulty depending on what properties the measure $d\mu$ is required to satisfy. We shall not deal with this problem. Our interest here lies on the nature of the relations between the sequence of polynomials $\{Q_n\}_{n \geq 0}$, the sequence of moments $\{\mu_n\}_{n \geq 0}$ and the two sequences $\{c_n\}_{n \geq 0}, \{\lambda_n\}_{n \geq 1}$. It turns out that these relations may be beautifully expressed by means of the theory of continued fractions. The corresponding identities are the contents of the following sequence of theorems which combine classical results of Jacobi, Rogers, Stieltjes and others. We start with a result that shows that each μ_n may actually be expressed as a polynomial in the c's and the λ's.

Theorem 4.5 *Let the sequence of polynomials $\{Q_n\}_{n\geq 0}$ be constructed by the recursions in (4.4.11) and the initial conditions in (4.4.12). Let μ_n be the coefficient of x^n in the continued fraction*

$$J(x; c, \lambda) = \cfrac{1}{1 - c_0 x - \cfrac{\lambda_1 x^2}{1 - c_1 x - \cfrac{\lambda_2 x^2}{1 - c_2 x - \cfrac{\lambda_3 x^2}{1 - c_3 x - \cdots}}}} \tag{4.4.15}$$

More precisely let the Taylor expansion of $J(x; c, \lambda)$ at $x = 0$ be

$$J(x; c, \lambda) = \sum_{n \geq 0} \mu_n x^n . \tag{4.4.16}$$

Then the polynomials Q_n do satisfy the orthogonality condition in (4.4.3) (2) with respect to the scalar product given by (4.4.14), (4.4.13) and the μ_n's given by (4.4.16).

The continued fraction $J(x; c, \lambda)$ is usually referred to as the *Jacobi continued fraction*. It reduces to the so called *Stieltjes continued fraction* when all the c_i's are equal to zero. The finite continued fraction

$$J^{(n)}(x; c, \lambda) = \cfrac{1}{1 - c_0 x - \cfrac{\lambda_1 x^2}{1 - c_1 x - \cfrac{\lambda_2 x^2}{1 - c_2 x - \cfrac{\cdots}{\cfrac{\cdots}{1 - c_{n-1} x - \cfrac{\lambda_n x^2}{1 - c_n x}}}}}}$$

is usually referred to as the nth *convergent* of $J(x; c, \lambda)$.

The next result relates $J^{(n)}(x; c, \lambda)$ to the polynomials Q_n's. To state it we need further notation. Given a polynomial $\phi(x; c_0, c_1, \ldots; \lambda_1, \lambda_2, \ldots)$ let $S\phi$ denote the polynomial obtained by replacing in ϕ each c_i by a c_{i+1} and each λ_i by a λ_{i+1}. We may refer to S itself as the *shift operator*.

This given we have:

Theorem 4.6

$$J^{(n)}(x; c, \lambda) = \frac{S Q_n^*(x)}{Q_{n+1}^*(x)} \tag{4.4.17}$$

where

$$Q_n^*(x) = x^n Q_n(\frac{1}{x}) \; , \qquad Q_{n+1}^*(x) = x^{n+1} Q_{n+1}(\frac{1}{x}) \; . \tag{4.4.18}$$

The next result is an identity which is customarily used to establish the convergence of $J^{(n)}(x; c, \lambda)$ to $J(x; c, \lambda)$ as $n \to \infty$.

Theorem 4.7

$$J^{(n)}(x; c, \lambda) - J^{(n-1)}(x; c, \lambda) = \frac{\lambda_1 \lambda_2 \cdots \lambda_n \, x^{2n}}{Q_n^*(x) Q_{n+1}^*(x)} \; . \tag{4.4.19}$$

An immediate corollary of (4.4.19) is that the rational function $J^{(n)}(x; c, \lambda)$ does converge to $J(x; c, \lambda)$ at least in the formal power series sense. In fact we see that the coefficient of x^n in the Taylor series expansion of $J^{(m)}(x; c, \lambda)$ is the same as that of $J(x; c, \lambda)$ itself as soon as $2m > n$. This shows that μ_n may be directly computed from $J^{(\frac{n-1}{2})}(x; c, \lambda)$ if n is odd and from $J^{(\frac{n}{2})}(x; c, \lambda)$ if n is even. In any case, we see that (4.4.17) defines it to be a polynomial in $c_0, c_1, \ldots, c_m$ and $\lambda_1, \lambda_2, \ldots, \lambda_m$ where m is the largest integer in $n/2$. Nevertheless, the computation of μ_n by means of one of the convergents $J^{(m)}(x; c, \lambda)$, requires (in view of (4.4.17)) the calculation of the Taylor series of the rational inverse of Q_{m+1}^*. To avoid this step Stieltjes devised an algorithm for the recursive computation of the moments μ_n. His result may be stated as follows:

Theorem 4.8 *Let* $H = \|h_{ij}\|_{i,j \geq 0}$ *be the lower triangular matrix determined by the double recursion*

$$h_{n,j} = h_{n-1,j-1} \lambda_j + h_{n-1,j} c_j + h_{n-1,j+1} \tag{4.4.20}$$

and the following initial conditions

$$h_{0,0} = 1 \, , \quad h_{n,j} = 0 \quad for \quad n < j \, . \tag{4.4.21}$$

Then

$$h_{nk} = \int_{-\infty}^{+\infty} x^n \, Q_k(x) \, d\mu = J(x; c, \lambda) \, Q_k^* \Big|_{x^{n+k}} \; . \tag{4.4.22}$$

In particular, the coefficient μ_n *in (4.4.16) can be computed by means of the identity*

$$\mu_n = h_{n,0} \, . \tag{4.4.23}$$

Some comments are in order here about the first equality in (4.4.22). Let $\{v_n\}_{n\geq 0}$ be a sequence of constants (with $v_0 = 0$) and let L_v be a linear functional on polynomials defined by setting

$$L_v(x^n) = v_n .$$

Suppose that the sequence Q_n given by the three term recursion in (4.4.11) with the initial conditions (4.4.12) is orthogonal with respect to the scalar product

$$\langle P, Q \rangle_v = L_v(PQ) .$$

Set

$$f_{n,k} = L_v(x^n Q_k) .$$

Then $Q_0 = 1$ and orthogonality gives that

$$f_{0,0} = 1, \quad f_{n,k} = 0 \quad \text{for} \quad n < k .$$

Moreover, using (4.4.11) we derive that

$$\begin{aligned}
f_{n+1,k} &= L_v(x^n(x Q_k)) \\
&= \lambda_k L_v(x^n Q_{k-1}) + c_k L_v(x^n Q_k) + L_v(x^n Q_{k+1}) \\
&= \lambda_k f_{n,k-1} + c_k f_{n,k} + f_{n,k+1} .
\end{aligned}$$

Thus the two matrices $\|f_{n,k}\|$ and $\|h_{n,k}\|$ must be identical. In particular, for $k = 0$ we deduce that

$$v_n = L_v(x^n) = h_{n,0} .$$

This shows that the three term recurrence uniquely determines the scalar product with respect to which the polynomials Q_k are to be orthogonal. In particular we see that the first equality in (4.4.22) is a immediate consequence of our definitions. Thus the essential part of Theorem 4.8 is given by the second equality in (4.4.22), which in particular states that the moment sequence μ_n is given by (4.4.16). In other words we see that Theorem 4.5 is a corollary of Theorem 4.7. Another useful consequence of (4.4.22) is the following result.

Theorem 4.9 *The matrix* $\|\frac{h_{n,k}}{(\lambda_1\lambda_2\cdots\lambda_k)}\|$ *is the inverse of the matrix* $\|a_{n,k}\|$ *of the coefficients of the polynomials* Q_n .

Proof Since $\{Q_n(x)\}_{n\geq 0}$ is an orthogonal basis, we have

$$\sum_{k=0}^{n} \frac{\langle y^n, Q_k(y)\rangle}{\langle Q_k, Q_k\rangle} Q_k(x) = x^n.$$

Expanding the polynomial $Q_k(x)$ we get

$$\sum_{k=0}^{n} \frac{\langle y^n, Q_k(y)\rangle}{\langle Q_k, Q_k\rangle} \sum_{s=0}^{k} a_{k,s}\, x^s = x^n.$$

Equating the coefficients of x^s on both sides of this identity we obtain

$$\sum_{k=0}^{n} \frac{\langle y^n, Q_k(y)\rangle}{\langle Q_k, Q_k\rangle} a_{k,s} = \chi(s=n). \qquad (4.4.24)$$

From (4.4.22) it follows that $\langle y^n, Q_k(y)\rangle = h_{n,k}$. But we also have $\langle Q_k, Q_k\rangle = \langle y^k, Q_k(y)\rangle = \lambda_1\lambda_2\cdots\lambda_k$. This is because a Motzkin path can reach height n from height 0 in n steps only proceeding by NORTHEAST steps all the way. Thus the identity in (4.4.24) can be written as

$$\sum_{k=0}^{n} \frac{h_{n,k}}{\lambda_1\lambda_2\cdots\lambda_k} a_{k,s} = \chi(s=n),$$

but that is exactly what the theorem states.   □

4.5   Moments and Motzkin Paths

Our goal in this section is to show that all the constructs we have dealt with in the previous section have combinatorial interpretations that make the identities stated in Theorems 4.5–4.9 visually self evident. To this end we shall deal here with a more general class of Motzkin paths. These are lattice paths that proceed by NORTHEAST, EAST and SOUTHEAST steps and remain throughout weakly above the x-axis without restrictions on the heights of the starting or ending points. For instance we give below a Motzkin path that starts at level 5 and ends at level 2:

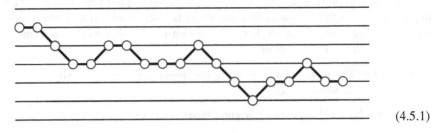

$$(4.5.1)$$

Here and after the collection of Motzkin paths that start at level a and end at level b will be denoted by $\Pi_{a,b}$. To such a path π we shall associate a word $w(\pi)$

by replacing (from left to right) each NORTHEAST edge by an a_i, each EAST edge by a c_i and each SOUTHEAST edge by a b_i, the subscript i giving the starting height of the edge. For instance, carrying this out on the path in (4.5.1) yields the word

$$c_5 b_5 b_4 c_3 a_3 c_4 b_4 a_3 b_4 c_3 c_3 a_3 b_4 b_3 b_2 a_1 c_2 a_2 b_3 c_2 .$$

Clearly, we can recover a path from its word. So in some contexts (see Sect. 4.8) it is important to regard a_i, b_i and c_i as sequences of noncommuting variables. Nevertheless, in the present context it simplifies our notation to allow these letters to commute, and we shall do so in the following as long as there is no loss. With this convention, we let L denote the formal power series given by the following summation

$$L = \sum_{\pi \in \Pi_{0,0}} x^{n(\pi)} w(\pi) , \tag{4.5.2}$$

where we let $n(\pi)$ denote here the total number of edges of π. We shall also require that the empty path contributes a term equal to 1 to the sum in (4.5.2). This given, a moment's reflection should reveal that this formal power series must satisfy the following identity

$$L = 1 + c_0 x\, L + x^2 a_0\, (SL)\, b_1\, L , \tag{4.5.3}$$

where S again denotes the *shift* operator that replaces each letter indexed by i by the same letter indexed by $i + 1$. Successive iterations of (4.5.3) then yield that L must be given by the continued fraction

$$L = L(x; a, b, c) = \cfrac{1}{1 - c_0 x - \cfrac{a_0 b_1 x^2}{1 - c_1 x - \cfrac{a_1 b_2 x^2}{1 - c_2 x - \cfrac{a_2 b_3 x^2}{1 - c_3 x - \cdots}}}} \tag{4.5.4}$$

Comparing with (4.4.15) we see that any specialization of the sequences $\{a_i\}$, $\{b_i\}$ that makes $a_{i-1} b_i = \lambda_i$ reduces $L(x; a, b, c)$ to $J(x; c, \lambda)$. Here and after we shall assume that this equality holds true. We thus immediately get:

Proposition 4.2 *The sequence μ_n defined by (4.4.16) has the following combinatorial interpretation*

$$\mu_n = \sum_{\pi \in \Pi_{0,0}(n)} w(\pi) \tag{4.5.5}$$

where $\Pi_{0,0}(n)$ denotes the collection of Moztkin paths of length n in $\Pi_{0,0}$.

Note further that if we set $a_{i-1} = b_i = c_i = 0$ for all $i > n$ then $L(x; a, b, c)$ reduces to $J^{(n)}(x; c, \lambda)$. Clearly, this substitution, wipes out from (4.5.2) all the terms produced by paths whose maximum height exceeds n. Consequently we must have

Proposition 4.3 (Viennot [25])

$$J^{(n)}(x; c, \lambda) = \sum_{\pi \in \Pi_{0,0}^{\leq n}} x^{n(\pi)} \, w(\pi) \tag{4.5.6}$$

where $\Pi_{0,0}^{\leq n}$ denotes the paths in $\Pi_{0,0}$ of maximum height not exceeding n.

Recalling the statement of Theorem 4.1 (especially part (2)) we deduce that $J^{(n)}(x; c, \lambda)$ must be closely related to the language of pyramids of monomers and dimers projecting on $[0, n]$ and summits m_0 or d_1. Note next that in the bijection giving Theorem 4.1 every monomer m_i comes from an EAST edge at height i and every dimer d_i may be made to correspond to a pair a_{i-1}, b_i of NORTHEAST and SOUTHEAST edges that respectively start at heights $i - 1$ and i. This means that if, in the above mentioned language of pyramids, we replace each m_i by xc_i and each d_i by $x^2 a_{i-1} b_i$ the result should be the truncated continued fraction $J^{(n)}(x; c, \lambda)$. We are thus led to the following result

Proposition 4.4 *We may write*

$$J^{(n)}(x; c, \lambda) = \frac{N_n}{D_n} \tag{4.5.7}$$

with

$$(i) \ D_n = \sum_{F \in \mathcal{F}([0,n], M)} (-1)^{|F|} \, w(F) \quad and \quad (ii) \ N_n = S D_{n-1} \tag{4.5.8}$$

where, $\mathcal{F}([0, n], M)$ denotes the collection of free sets of letters in the alphabet of monomer and dimers projecting on the interval $[0, n]$ and $w(F)$ is the word obtained from F by setting $m_i = xc_i$ and $d_i = x^2 a_{i-1} b_i$.

Proof Theorem 4.3 with A the alphabet of monomers and dimers projecting on $[0, n]$ and $B = \{m_0, d_1\}$ yields that we must have (4.5.7) with D_n as given by (4.5.8) (i) and

$$N_n = \sum_{F \in \mathcal{F}([1,n], M)} (-1)^{|F|} \, w(F) \tag{4.5.9}$$

where $\mathcal{F}([1, n], M)$ denotes the collection of free sets of monomers and dimers projecting on $[1, n]$. However, since every free set projecting on $[1, n]$ is an image by S of a free set projecting on $[0, n-1]$, we must also have (4.5.8) (ii) as asserted.

□

We are now ready for the following result.

The proof of Theorem 4.6 Note that a free set F in (4.5.8) (i) may either project on $[0, n-1]$ or may be written in the form $F = F_1 \cup \{m_n\}$ with F_1 projecting on $[0, n-1]$ or in the form $F = F_2 \cup \{d_n\}$ with F_2 projecting on $[0, n-2]$. This yields the recursion

$$D_n = D_{n-1}(1 - w(m_n)) - D_{n-2}w(d_n) \tag{4.5.10}$$

and since we are supposed to let $w(m_i) = xc_i$ and $w(d_i) = x^2 a_{i-1} b_i$, we see that (4.5.10) is none other than

$$D_n = D_{n-1}(1 - xc_n) - D_{n-2}x^2 \lambda_n . \tag{4.5.11}$$

On the other hand the definitions in (4.4.18) combined with the three term recursion in (4.4.11) yield that

$$Q_{n+1}^* = (1 - xc_n)Q_n^* - x^2 \lambda_n Q_{n-1}^* .$$

Now by definition

$$J^{(0)}(x; c, \lambda) = \frac{1}{1 - c_0 x} \quad \text{and} \quad J^{(1)}(x; c, \lambda) = \frac{1 - c_1 x}{(1 - c_0 x)(1 - c_1 x) - \lambda_1 x^2}$$

which gives

$$D_0 = 1 - c_0 x = Q_1^* \quad \text{and} \quad D_1 = (1 - c_0 x)(1 - c_1 x) - \lambda_1 x^2 = Q_2^* .$$

Thus the two sequences $\{D_n\}_{n \geq 0}$ and $\{Q_{n+1}^*\}_{n \geq 0}$ must be identical since they satisfy the same recursion and the same initial conditions. This completes the proof of (4.4.17)

Before proceeding with the proof of the identities in (4.4.19) and (4.4.22) we should note that our Motzkin paths $\Pi_{0,k}$ provide us with an immediate combinatorial interpretation of the matrix entries $h_{n,k}$.

To be precise we have

Proposition 4.5

$$h_{n,k} = \sum_{\pi \in \Pi_{0,k}(n)} w(\pi) \Big|_{\substack{a_n=\lambda_{n+1} \\ b_n=1}} \tag{4.5.12}$$

where $\Pi_{0,k}(n)$ denotes the collection of paths that go from height 0 to height k in n steps.

Proof Note that every path in $\Pi_{0,k}(n)$ must come either from height $k - 1$ by a NORTHEAST step or from height k by an EAST step or from height $k + 1$ by a SOUTHEAST step. Letting $\tilde{h}_{n,k}$ denote, for a moment, the right hand side of (4.5.12), this observation leads to the recursion

$$\tilde{h}_{n,k} = \tilde{h}_{n-1,k-1}a_{k-1} + \tilde{h}_{n-1,k}c_k + \tilde{h}_{n-1,k+1}b_{k+1} \, ,$$

which reduces to (4.4.20) when we set $a_n = \lambda_{n+1}$ and $b_n = 1$ for all $n \geq 0$. Note further that every path with n edges remains below height n and reaches that height only when it consists totally of NORTHEAST steps. Thus when $n < k$, the right hand side of (4.5.12) reduces to an empty sum, which forces $\tilde{h}_{n,k} = 0$. Since trivially we have also $\tilde{h}_{0,0} = 1$, we see that $h_{n,k}$ and $\tilde{h}_{n,k}$ do satisfy the same double recursion and the same initial conditions. Thus (4.5.12) must hold true as asserted.   □

To complete our combinatorial derivation of the identities of the last section we need an auxiliary result.

Proposition 4.6 *(Viennot [25])*

$$\frac{b_1 b_2 \cdots b_n x^n}{D_n} = \sum_{\pi \in \Pi_{n,0}^{\leq n}} x^{n(\pi)} w(\pi) \tag{4.5.13}$$

where $\Pi_{n,0}^{\leq n}$ denotes the set of Motzkin paths that go from height n to height 0 always remaining weakly below height n.

Proof We start by observing that every monomer-dimer heap α on the interval $[0, n]$ may be uniquely decomposed into a product $\alpha = \alpha_0 \alpha_1 \cdots \alpha_{n-1} \alpha_n$ with α_i a pyramid projecting on $[i, n]$ with summit a monomer m_i or a dimer d_{i+1}. Of course, we do not exclude the possibility that some of the α_i may be empty here. To show this, a few pictures should suffice. For instance we give below a monomer dimer heap which projects on the interval $[0, 8]$. Actually to properly see it we must turn the picture 90° clockwise.

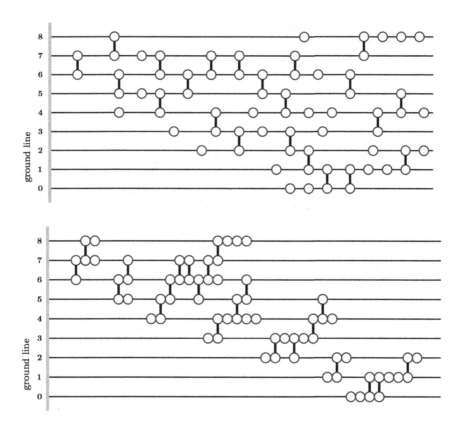

Starting from level zero we locate the leftmost piece (a monomer here). Then we group together the pieces that would have to move if we dragged that leftmost piece all the way to the right. Repeating for each level the process we applied to level 1 we obtain the first diagram of the above display.

The second diagram exhibits the resulting groups. To facilitate the construction of each individual group, we maximally compressed each group and inserted spaces in between.

To proceed with our proof, we now use the bijection of Theorem 4.1 and convert each of these heaps back into a Motzkin path. We see that α_i will necessarily convert into a path π_i which starts at level i and ends at level i remaining weakly above that level throughout. Finally note that inserting between π_i and π_{i-1} (for $i = n, n - 1, \ldots, 2, 1$) a SOUTHEAST step b_i we necessarily produce a Motzkin path π, which starts at level n ends at level 0 remaining throughout weakly below level n. We can see then that (with the substitutions $m_i = c_1 x$ and $d_i = a_{i-1} b_i x^2$) we shall have

$$x^{n(\pi)} w(\pi) = x^{n(\pi_n)} w(\pi_n) b_n x x^{n(\pi_{n-1})} w(\pi_{n-1}) b_{n-1} x \cdots$$

$$\cdots x^{n(\pi_2)} w(\pi_2) b_2 x x^{n(\pi_1)} w(\pi_1) b_1 x x^{n(\pi_0)} w(\pi_0)$$

$$= w(\alpha_n) b_n x w(\alpha_{n-1}) b_{n-1} x \cdots w(\alpha_2) b_2 x w(\alpha_1) b_1 x w(\alpha_0) \ .$$

$$(4.5.14)$$

Now, from the definition (4.5.8) (i) and the Cartier–Foata Theorem 4.2, we see that the rational function $1/D_n$ gives the sum of the weights of the monomer dimer heaps projecting on $[0, n]$. Note further that the construction illustrated above maps this collection of heaps onto paths in $\Pi_{n,0}^{\leq n}$ *bijectively*. To see this we only need a single picture. We have illustrated in Fig. (4.5.15) a path π in $\Pi_{10,0}^{\leq 10}$.

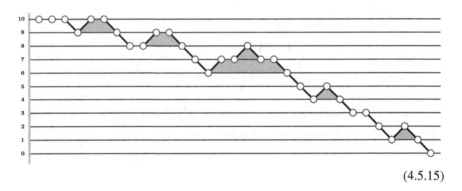

$$(4.5.15)$$

The shaded areas are precisely the regions left in the shade if the sun is shining from the point at $-\infty$ on the x-axis (as if some SE edges are transparent). We see that, removing the transparent edges, the path π breaks up into smaller Motzkin paths $\pi_n, \pi_{n-1}, \ldots, \pi_1, \pi_0$ where π_i starts and ends at height i and remains above that height throughout. Of course, as we see in the example above, some of the π_i's could very well be empty. Nevertheless we can convert each π_i into a heap α_i by the bijection of Theorem 4.1. This produces a heap $\alpha = \alpha_n \alpha_{n-1} \cdots \alpha_1 \alpha_0$ projecting on $[0, n]$. Now it is easy to see that α is precisely the heap that would produce π by the construction illustrated above. Thus we must conclude that (4.5.13) is simply obtained by summing the relation in (4.5.14) as α varies over all heaps projecting on $[0, n]$. This completes our proof.   □

Remark 4.1 Note that by reversing the order of the paths π_i in (4.5.14) and replacing the intermediate SOUTHEAST edges by NORTHEAST edges we obtain a Motzkin path π' which goes from level 0 to level n remaining throughout below level n. This construction replaces (4.5.14) by the identity

$$x^{n(\pi')} w(\pi') = x^{n(\pi_0')} w(\pi_0) a_0 x x^{n(\pi_1')} w(\pi_1) a_1 x x^{n(\pi_2')} w(\pi_2) a_2 x \cdots$$

$$\cdots x^{n(\pi_{n-1}')} w(\pi_{n-1}) a_{n-1} x x^{n(\pi_n')} w(\pi_n)$$

$$= w(\alpha_0) a_0 x w(\alpha_1) a_1 x w(\alpha_2) a_2 x \cdots w(\alpha_{n-1}) a_{n-1} x w(\alpha_n) .$$

$$(4.5.16)$$

It goes without saying that this observation shows that we also have a bijection between Motzkin paths that go from height 0 to height n remaining weakly below height n and monomer dimer heaps on $[0, n]$. Thus, denoting that collection of Motzkin paths by $\Pi_{0,n}^{\leq n}$, we see that summing (4.5.16) over all $\pi' \in \Pi_{0,n}^{\leq n}$ yields the companion identity

$$\frac{a_0 a_1 \cdots a_{n-1} x^n}{D_n} = \sum_{\pi \in \Pi_{0,n}^{\leq n}} x^{n(\pi)} w(\pi) . \qquad (4.5.17)$$

We are now in a position to prove and (4.4.19) and (4.4.22). Note first that our Motzkin path interpretation of $J^{(n)}(x; c, \lambda)$ immediately yields that

$$J^{(n)}(x; c, \lambda) - J^{(n-1)}(x; c, \lambda) = \sum_{\pi \in \Pi_{0,0}^{(\max=n)}} x^{n(\pi)} w(\pi) , \qquad (4.5.18)$$

where $\Pi_{0,0}^{(\max=n)}$ denotes the collection of Motzkin paths in $\Pi_{0,0}$ whose maximum height is equal to n. Now a path $\pi \in \Pi_{0,0}^{(\max=n)}$ may be uniquely factorized in to a path $\pi_1 \in \Pi_{0,n}^{\leq n}$ followed by a SOUTHEAST step b_n and a path $\pi_2 \in \Pi_{n-1,0}^{\leq n-1}$. A moment's reflection leads us to conclude that this factorization yields the identity

$$J^{(n)}(x; c, \lambda) - J^{(n-1)}(x; c, \lambda) \qquad (4.5.19)$$

$$= \frac{a_0 a_1 \cdots a_{n-1} x^n}{D_n} \times b_n x \times \frac{b_{n-1} b_{n-2} \cdots b_2 b_1 x^{n-1}}{D_{n-1}} .$$

Indeed, summing over all possible π_1 and using (4.5.17) yields the first factor, and summing over all possible π_2 and using (4.5.13) (with n replaced by $n - 1$) yields the last factor. Now we have shown in the proof of Theorem 4.6 that $D_n = Q_{n+1}^*$ and $D_{n-1} = Q_n^*$. Thus making these substitutions and setting $a_{i-1} b_i = \lambda_i$ we see that (4.5.19) reduces to (4.4.19).

To show (4.4.22) we start by observing that the difference

$$J(x; c, \lambda) - J^{(n-1)}(x; c, \lambda)$$

may be given the combinatorial interpretation

$$J(x; c, \lambda) - J^{(n-1)}(x; c, \lambda) = \sum_{\pi \in \Pi_{0,0}^{\max \geq n}} x^{n(\pi)} w(\pi) , \qquad (4.5.20)$$

where $\Pi_{0,0}^{\max \geq n}$ denotes the collection of all Motzkin paths in $\Pi_{0,0}$ which reach heights weakly above n. Now any $\pi \in \Pi_{0,0}^{\max \geq n}$ may be uniquely factorized into a path $\pi_1 \in \Pi_{0,n}$ followed by a SOUTHEAST edge and a path $\pi_2 \in \Pi_{n-1,0}^{(\leq n-1)}$. The same reasoning that led us to (4.5.19) now yields

$$J(x; c, \lambda) - J^{n-1}(x; c, \lambda) = \left(\sum_{\pi \in \Pi_{0,n}} x^{n(\pi)} w(\pi) \right) \times b_n x \times \frac{b_{n-1} \cdots b_2 b_1 x^{n-1}}{D_{n-1}} .$$

Multiplying both sides by D_{n-1} and using (4.5.7) (with n replaced by $n - 1$) we obtain the identity

$$b_n b_{n-1} \cdots b_1 x^n \sum_{\pi \in \Pi_{0,n}} x^{n(\pi)} w(\pi) = J(x; c, \lambda) D_{n-1}(x) - N_{n-1}(x) .$$

$$(4.5.21)$$

We now replace n by k, D_{n-1} and N_{n-1} respectively by Q_k^* and SQ_{k-1}^* and derive that

$$b_1 b_2 \cdots b_k x^k \sum_{\pi \in \Pi_{0,k}} x^{n(\pi)} w(\pi) = J(x; c, \lambda) Q_k^*(x) - SQ_{k-1}^*(x) . \qquad (4.5.22)$$

Equating coefficients of x^{k+n} and using our combinatorial interpretation (4.5.12) for the matrix elements $h_{n,k}$, we finally get that, with the specializations $a_n = \lambda_{n+1}$ and $b_n = 1$ (for all $n \geq 0$) we must have

$$h_{n,k} = J(x; c, \lambda) Q_k^*(x) \Big|_{x^{k+n}} . \qquad (4.5.23)$$

The term $SQ_{k-1}^*(x)$ in (4.5.22) contributes nothing to (4.5.23) for the simple reason that it is a polynomial of degree $k - 1$. This shows the equality of the first and last member of (4.4.22). To complete the proof of (4.4.22) we need only express the right hand side of (4.5.23) in terms of the polynomial Q_k. To this end note that if $Q_k = \sum_{s=0}^k a_{ks} x^s$ then from (4.4.16) we get that

$$J(x; c, \lambda) Q_k^*(x) = \left(\sum_{m \geq 0} \mu_m x^m \right) \left(\sum_{s=0}^k a_{ks} x^{k-s} \right) .$$

Equating coefficients of x^{k+n} gives

$$J(x; c, \lambda) \, Q_k^*(x) \Big|_{x^{n+k}} = \sum_{s=0}^{k} a_{ks} \, \mu_{n+s} \, . \qquad (4.5.24)$$

So if the μ_n's are the moments of a measure μ we may rewrite this result in the form

$$J(x; c, \lambda) \, Q_k^*(x) \Big|_{x^{n+k}} = \int_{-\infty}^{+\infty} Q_k(x) \, x^n \, d\mu \, . \qquad (4.5.25)$$

This proves the last equality in (4.4.23).

If we do not want to make use of a measure, we can work with the bilinear form defined by (4.4.13) and (4.4.14) and rewrite (4.5.24) in the form

$$J(x; c, \lambda) \, Q_k^*(x) \Big|_{x^{n+k}} = \langle \, Q_k \, , \, x^n \, \rangle_\mu \, . \qquad (4.5.26)$$

Since $h_{n,k} = 0$ for $n < k$, from (4.5.24) we deduce that

$$\langle \, Q_k \, , \, x^n \, \rangle_\mu = 0 \qquad \text{(for } n < k) \, .$$

In particular, we must also have

$$\langle \, Q_k \, , \, Q_n \, \rangle_\mu = 0 \qquad \text{(for } n < k) \, . \qquad (4.5.27)$$

In other words the Q_n's are orthogonal precisely as asserted at the end of Theorem 4.5. This completes our proof of the identities of Sect. 4.4.

The identity in (4.5.21) has an extension that is worth mentioning here.

Proposition 4.7 (Viennot [25]) *For $0 \le h \le k$ we have*

$$D_{h-1} \left(J(x; c, \lambda) D_{k-1} - N_{k-1} \right) = \lambda_1 \lambda_2 \cdots \lambda_h \, x^{h+k} \sum_{\pi \in \Pi_{h,k}} x^{n(\pi)} \, w(\pi) \, . \qquad (4.5.28)$$

Proof We start with (4.5.21) written for $n = k$

$$J(x; c, \lambda) D_{k-1}(x) - N_{k-1}(x) = b_k b_{k-1} \cdots b_1 x^k \sum_{\pi \in \Pi_{0,k}} x^{n(\pi)} \, w(\pi) \, . \qquad (4.5.29)$$

Note that each path $\pi \in \Pi_{0,k}$ has a unique decomposition into a path $\pi_1 \in \Pi_{0,h-1}^{\le h-1}$ followed by a NORTHEAST step, followed by a path $\pi_2 \in \Pi_{h,k}$. Indeed, the cutting point for π_1 is precisely before the edge where π reaches for the first time height h.

This gives

$$w(\pi) = w(\pi_1) \, a_{h-1} x \, w(\pi_2) .$$

Summing over all possible choices of π_1 and π_2 and using (4.5.17) (for $n = h - 1$) gives the identity

$$\sum_{\pi \in \Pi_{0,k}} x^{n(\pi)} \, w(\pi) = \frac{a_0 a_1 \cdots a_{h-2} x^{h-1}}{D_{h-1}} \, a_{h-1} x \sum_{\pi \in \Pi_{h,k}} x^{n(\pi)} \, w(\pi) .$$

Substituting in (4.5.29) and setting $a_i = \lambda_i$ and $b_i = 1$ gives

$$J(x; c, \lambda) D_{k-1}(x) - N_{k-1}(x) = \frac{\lambda_1 \lambda_2 \cdots \lambda_h \, x^{h+k}}{D_{h-1}} \sum_{\pi \in \Pi_{h,k}} x^{n(\pi)} \, w(\pi) ,$$

and (4.5.28) follows upon multiplying both sides by D_{h-1}.   □

Notice that since D_{h-1} and N_{k-1} are polynomials of degrees h and $k - 1$ respectively, equating coefficients of x^{h+k+n} of both sides of (4.5.28) gives

$$D_{h-1} \, J(x; c, \lambda) D_{k-1} \Big|_{x^{h+k+n}} = \lambda_1 \lambda_2 \cdots \lambda_h \sum_{\pi \in \Pi_{h,k}(n)} w(\pi) , \qquad (4.5.30)$$

where $\Pi_{h,k}(n)$ denotes the collection of Motzkin paths that go from height h to height k in n steps. Recalling that $D_{h-1} = Q_h^*$ and $D_{k-1} = Q_k^*$ and imitating our derivation of (4.5.25) from (4.5.23) we are finally led to the identity

$$\int_{-\infty}^{+\infty} Q_h(x) Q_k(x) \, x^n \, d\mu = \lambda_1 \lambda_2 \cdots \lambda_h \sum_{\pi \in \Pi_{h,k}(n)} w(\pi) .$$

Our combinatorial setting can be used to establish further identities. Note that combining (4.4.22) and (4.5.12) we may write

$$\int_{-\infty}^{+\infty} x^n \, Q_k \, d\mu = \sum_{\pi \in \Pi_{0,k}(n)} w(\pi) \Big|_{\substack{a_{i-1}=\lambda_i \\ b_i = 1}} . \qquad (4.5.31)$$

This identity yields us a combinatorial way of deriving explicit expressions for λ_n and c_n in terms of the moment sequence $\{\mu_n\}$.

Proposition 4.8

$$(a) \; \lambda_n = \frac{d_n d_{n-2}}{d_{n-1}^2} , \qquad (b) \; c_n = \frac{\chi_n}{d_n} - \frac{\chi_{n-1}}{d_{n-1}} \qquad (4.5.32)$$

where

$$d_n = \det \begin{pmatrix} \mu_0 & \mu_1 & \cdots & \mu_n \\ \mu_1 & \mu_2 & \cdots & \mu_{n+1} \\ \cdots & \cdots & \cdots & \cdots \\ \mu_{n-1} & \mu_n & \cdots & \mu_{2n-1} \\ \mu_n & \mu_{n+1} & \cdots & \mu_{2n} \end{pmatrix}, \tag{4.5.33}$$

$$\chi_n = \det \begin{pmatrix} \mu_0 & \mu_1 & \cdots & \mu_n \\ \mu_1 & \mu_2 & \cdots & \mu_{n+1} \\ \cdots & \cdots & \cdots & \cdots \\ \mu_{n-1} & \mu_n & \cdots & \mu_{2n-1} \\ \mu_{n+1} & \mu_{n+2} & \cdots & \mu_{2n+1} \end{pmatrix}.$$

Proof Since there is only one way for a Motzkin path to reach height n from height 0 in n steps (NORTHEAST all the way), formula (4.5.31) for $k = n$ reduces to

$$\int_{-\infty}^{+\infty} x^n \, Q_n \, d\mu = a_0 a_1 \cdots a_{n-1} \Big|_{a_{i-1}=\lambda_i} = \lambda_1 \lambda_2 \cdots \lambda_n . \tag{4.5.34}$$

On the other hand, using (4.4.4) we get

$$\int_{-\infty}^{+\infty} x^n \, Q_n \, d\mu = \frac{d_n}{d_{n-1}} . \tag{4.5.35}$$

Combining with (4.5.34) we deduce that

$$\frac{d_n}{d_{n-1}} = \lambda_1 \lambda_2 \cdots \lambda_n , \tag{4.5.36}$$

and (4.5.32) (a) immediately follows. $\qquad\qquad\square$

Notice next that (4.4.4) and (4.5.33) yield that

$$\int_{-\infty}^{+\infty} x^{n+1} \, Q_n \, d\mu = \frac{1}{d_{n-1}} \chi_n . \tag{4.5.37}$$

On the other hand a Motzkin path can reach height n from height 0 in $n+1$ steps if and only if it takes i successive NORTHEAST steps followed by a single EAST step and then finish up with $n - i$ successive NORTHEAST steps (for $i = 0, 1, \ldots, n$). This gives

$$\sum_{\pi \in \Pi_{0,k}(n)} w(\pi) \Big|_{\substack{a_{i-1}=\lambda_i \\ b_i=1}} = (c_0 + c_1 + \cdots + c_n) \lambda_1 \lambda_2 \cdots \lambda_n . \tag{4.5.38}$$

Combining with (4.5.31), (4.5.36) and (4.5.37) we get

$$\chi_n = (c_0 + c_1 + \cdots + c_n) \, d_n \qquad (4.5.39)$$

which implies (4.5.32) (b) as desired.

4.6 Chebyshev Polynomials

The simplest family of orthogonal polynomials, from the combinatorial point of view developed in the previous sections, should be one that corresponds to Dyck paths (that is $c_n = 0$) with constant λ_n. It happens that the Chebyshev polynomials of the first and second kind come pretty near to this. Let us recall that the latter are the polynomials $\{\alpha_n(x)\}_n$ defined by setting

$$\alpha_n(\cos\theta) = \cos n\theta \qquad \text{(for } n \geq 0)\,, \qquad (4.6.1)$$

while those of the second kind may be implicitly defined as the $\{\beta_n(x)\}_n$ satisfying

$$\beta_n(\cos\theta) = \frac{\sin(n+1)\theta}{\sin\theta} \qquad \text{(for } n \geq 0)\,. \qquad (4.6.2)$$

Note that the trigonometric identities

$$\begin{aligned}
&(a) \quad \cos(n+1)\theta + \cos(n-1)\theta = 2\cos\theta\cos n\theta \\
&(b) \quad \sin(n+2)\theta + \sin n\theta = 2\cos\theta\sin(n+1)\theta
\end{aligned} \qquad (4.6.3)$$

immediately yield the recursions

$$\begin{aligned}
&(a) \quad \alpha_{n+1}(x) = 2x\,\alpha_n(x) - \alpha_{n-1}(x)\,, \\
&(b) \quad \beta_{n+1}(x) = 2x\,\beta_n(x) - \beta_{n-1}(x)\,.
\end{aligned} \qquad \text{(for } n \geq 1) \qquad (4.6.4)$$

Computing directly from the definitions we also get that

$$\begin{aligned}
&(a) \quad \alpha_0(x) = 1\,, \quad \alpha_1(x) = x \\
&(b) \quad \beta_0(x) = 1\,, \quad \beta_1(x) = 2x
\end{aligned} \qquad (4.6.5)$$

We see then that (4.6.4) (a) and (4.6.5) (a) determine $\alpha_n(x)$ has a polynomial in x of degree n and leading coefficient 2^{n-1}. Similarly, (4.6.4) (b) and (4.6.5) (b) yield that $\beta_n(x)$ is a polynomial degree n and leading coefficient 2^n. To get sequences of polynomials with leading term x^n as required by (4.4.3) (1) we must rescale. We shall thus set

$$
\begin{aligned}
(a)\ & A_n(x) = \alpha_n(x)/2^{n-1}\ , \\
(b)\ & B_n(x) = \beta_n(x)/2^n\ .
\end{aligned}
\qquad \text{(for } n \geq 1) \qquad (4.6.6)
$$

This has the effect of changing (4.6.4) (a) to

$$
\begin{aligned}
A_2(x) &= x\, A_1(x) - \tfrac{1}{2} A_0(x) \ \text{and} \\
A_{n+1}(x) &= x\, A_n(x) - \tfrac{1}{4} A_{n-1}(x) \ \text{(for } n \geq 2)
\end{aligned}
\qquad (4.6.7)
$$

with

$$
A_0(x) = 1 \quad \text{and} \quad A_1(x) = x \ . \qquad (4.6.8)
$$

While (4.6.4) (b) becomes

$$
B_{n+1}(x) = x\, B_n(x) - \tfrac{1}{4} B_{n-1}(x)\ , \qquad \text{(for } n \geq 1) \qquad (4.6.9)
$$

with

$$
B_0(x) = 1 \quad \text{and} \quad B_1(x) = x \ . \qquad (4.6.10)
$$

It will also be convenient to set

$$
A_n^*(x) = x^n A_n \left(\frac{1}{x}\right) \quad \text{and} \quad B_n^*(x) = x^n B_n \left(\frac{1}{x}\right) \ . \qquad (4.6.11)
$$

This given, comparing with (4.4.11) we immediately derive from Theorems 4.5 and 4.6 that

Theorem 4.10

$$
\lim_{n \to \infty} \frac{B_n^*(x)}{A_{n+1}^*(x)} = \cfrac{1}{1 - \cfrac{x^2/2}{1 - \cfrac{x^2/4}{1 - \cfrac{x^2/4}{1 - \cfrac{x^2/4}{1 - \cdots}}}}} \qquad (4.6.12)
$$

and

$$\lim_{n \to \infty} \frac{B_n^*(x)}{B_{n+1}^*(x)} = \cfrac{1}{1 - \cfrac{x^2/4}{1 - \cfrac{x^2/4}{1 - \cfrac{x^2/4}{1 - \cfrac{x^2/4}{1 - \cdots}}}}} \qquad (4.6.13)$$

Moreover, the α_n's and β_n's are orthogonal with respect to measures μ^a and μ^b with moments given by

$$\mu_{2n}^a = \frac{1}{2^{2n-1}} \frac{1}{n+1} \binom{2n}{n}, \quad \mu_{2n-1}^a = 0 \qquad (4.6.14)$$

and

$$\mu_{2n}^b = \frac{1}{2^{2n}} \frac{1}{n+1} \binom{2n}{n}, \quad \mu_{2n-1}^b = 0. \qquad (4.6.15)$$

Proof We only need to explain the numerator $B_n^*(x)$ on the left hand side of (4.6.12). To this end note that our combinatorial setting gives us that

$$A_{n+1}^*(x) = \sum_{F \in \mathcal{F}([0,n],M)} (-1)^{|F|} w(F) \Big|_{d_i = \lambda_i x^2} \qquad (4.6.16)$$

with $\lambda_1 = 1/2$ and $\lambda_n = 1/4$ for all $n \geq 2$. On the other hand (4.5.8) (ii) gives that the numerator in the left hand side of (4.6.12) is given by

$$SD_{n-1} = \sum_{F \in \mathcal{F}([1,n],M)} (-1)^{|F|} w(F) \Big|_{d_i = \lambda_i x^2} \qquad (4.6.17)$$

and this is equal to $B_n^*(x)$ since in the interval $[1, n]$ we can't place the dimer d_1 and we can set $d_i = x^2/4$ (for all i) on the right hand side of (4.6.17).   □

Remark 5.1 Note that (4.6.16) gives that the polynomials $A_n^*(x)$ are all functions of x^2, clearly the same holds true for all the $B_n^*(x)$. The relations in (4.6.11) then imply that $\alpha_n(x)$ and $\beta_n(x)$ are even functions of x for even n and odd functions for n odd. This can also be easily deduced from the recursions in (4.6.4). It is also seen that the combinatorial interpretation

$$B_n^*(x) = \sum_{F \in \mathcal{F}([0,n-1],M)} (-1)^{|F|} w(F) \Big|_{d_i = \lambda_i x^2}$$

yields us that

$$B_n^*(x) = \sum_{k=0}^{n} \binom{n-k}{k} (-1)^k \left(\frac{x^2}{4}\right)^k .$$

This is because there are precisely $\binom{n-k}{k}$ ways of placing k nonoverlapping dimers on $[0, n-1]$. Using (4.6.11) and (4.6.6) we are thus led to the explicit formula

$$\beta_n(x) = \sum_{k=0}^{n} \binom{n-k}{k} (-1)^k (2x)^{n-2k} . \tag{4.6.18}$$

Now this is neater than the formula yielded by the trigonometric identity

$$\cos(n+1)\theta + i \sin(n+1)\theta = (\cos\theta + i \sin\theta)^{n+1} .$$

In fact, equating imaginary parts we get

$$\sin(n+1)\theta = \sum_{m=0}^{n+1} \binom{n+1}{2m+1} (-1)^m \cos^{n-2m}\theta \sin^{2m+1}\theta$$

which using (4.6.2) translates into

$$\beta_n(x) = \sum_{m=0}^{n+1} \binom{n+1}{2m+1} (-1)^m x^{n-2m} (1-x^2)^m . \tag{4.6.19}$$

We should note that equating coefficients of x^{n-2k} in (4.6.18) and (4.6.19) leads to the binomial identity

$$\binom{n-k}{k} 2^{n-2k} = \sum_{s} \binom{n+1}{2k+2s+1} \binom{k+s}{s} . \tag{4.6.20}$$

The measures μ^a and μ^b corresponding to the two sequences $\{\alpha_n(x)\}_n$ and $\{\beta_n(x)\}_n$ are easily obtained in this case. Indeed, since

$$\frac{1}{\pi} \int_0^\pi \cos(n\theta) \cos(m\theta) \, d\theta = \begin{cases} 1 & \text{if } n = m = 0 \\ \frac{1}{2} & \text{if } n = m \geq 1 \\ 0 & \text{otherwise} \end{cases}$$

we derive from (4.6.1) (by the substitution $x = \cos\theta$) that

$$\frac{1}{\pi} \int_{-1}^{1} \alpha_n(x)\, \alpha_m(x)\, \frac{dx}{\sqrt{1-x^2}} = \begin{cases} 1 & \text{if } n = m = 0 \\ \frac{1}{2} & \text{if } n = m \geq 1 \\ 0 & \text{otherwise} . \end{cases}$$

Similarly, from

$$\frac{2}{\pi} \int_{0}^{\pi} \sin(n+1)\theta\, \sin(m+1)\theta\, d\theta = \begin{cases} 1 & \text{if } n = m \\ 0 & \text{otherwise} \end{cases}$$

making the substitution $x = \cos\theta$ and noting that $\sin^2\theta\, d\theta = -\sqrt{1-x^2}\, dx$ we get

$$\frac{2}{\pi} \int_{-1}^{1} \beta_n(x)\, \beta_m(x)\sqrt{1-x^2} dx = \begin{cases} 1 & \text{if } n = m \\ 0 & \text{otherwise} . \end{cases}$$

We thus obtain that μ^a and μ^b are both unit measures concentrated on the interval $[-1, 1]$ with

$$d\mu^a = \frac{1}{\pi} \frac{1}{\sqrt{1-x^2}}\, dx \quad \text{and} \quad d\mu^b = \frac{2}{\pi}\sqrt{1-x^2}\, dx .$$

In particular, comparing with (4.6.14) and (4.6.15) we obtain a combinatorial interpretation for the identities

$$\frac{1}{\pi} \int_{-1}^{1} x^{2n}\, \frac{dx}{\sqrt{1-x^2}} = \frac{1}{2^{2n-1}} \frac{1}{n+1} \binom{2n}{n} \quad \text{and}$$

$$\frac{2}{\pi} \int_{-1}^{1} x^{2n}\sqrt{1-x^2}\, dx = \frac{1}{2^{2n}} \frac{1}{n+1} \binom{2n}{n} .$$

4.7 The Rogers–Ramanujan Continued Fraction

The next simplest example is obtained by setting $a_n = q^{n+1}$, $b_n = 1$ and $c_n = 0$. With this choice we are again left with summing only over Dyck paths. It develops that both the moment sequence and the corresponding polynomials D_n and N_n have interesting combinatorial interpretations.

Theorem 4.11 *For a given Dyck path π let $a(\pi)$ denote number of lattice squares weakly below π and weakly above the x-axis, and let $\mathcal{D}_n$ denote the collection of*

Dyck paths of length 2n. Then we have

$$J(x; q) = \cfrac{1}{1 - \cfrac{qx^2}{1 - \cfrac{q^2 x^2}{1 - \cfrac{q^3 x^2}{1 - \cdots}}}} = \sum_{n \geq 0} c_n(q) \, q^n \, x^{2n} \, . \tag{4.7.1}$$

where

$$c_n(q) = \sum_{\pi \in \mathcal{D}_n} q^{a(\pi)} \, . \tag{4.7.2}$$

Proof The moments of the measure corresponding to the continued fraction on the left hand side of (4.7.1), according to (4.5.5), are given by the identity

$$\mu_n = \sum_{\pi \in \Pi_{0,0}} w(\pi) \, \Big|_{\lambda_i = q^i} \, . \tag{4.7.3}$$

A look at Fig. 4.7.4 should make it obvious that the contribution of a NORTHEAST edge e of a Dyck path π to

$$w(\pi) \, \Big|_{\substack{c_i = 0 \\ \lambda_i = q^i}}$$

is the factor q^{m+1}, where m is the number of lattice squares southeast of e that are weakly below π and weakly above the x-axis. Since setting $c_i = 0$ kills all but the summands corresponding to Dyck paths we see that the sum on the right hand side of (4.7.3) must evaluate to $q^n c_n(q)$ and thus (4.7.1) must hold true as asserted. $\qquad \square$

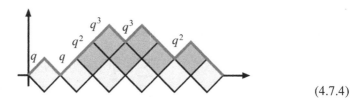

$$\tag{4.7.4}$$

Let now $D_n(x; q)$ denote the denominator of the nth convergent of the continued fraction $J(x; q)$ in (4.7.1). Note that the recursion in (4.5.11) in this case reduces to

$$D_n(x; q) = D_{n-1}(x; q) - x^2 q^n D_{n-2}(x; q) \, . \tag{4.7.5}$$

Note further that the dimers $d_{i_1}, d_{i_2}, \ldots, d_{i_k}$ do not overlap if and only if

$$i_1 + 2 \le i_2, \ i_2 + 2 \le i_3, \ \cdots, \ i_{k-1} + 2 \le i_k$$

In other words there is an obvious bijection between k-sets of nonoverlapping dimers and k-part partitions with part differences greater or equal to 2. For convenience let $\Delta_2(k)$ denote this collection of partitions and Δ_2 denote the union of all the $\Delta_2(k)$. Now it is well known and easy to show that we have

$$\sum_{\delta \in \Delta_2(k)} q^{|\delta|} = \frac{q^{k^2}}{(1-q)(1-q^2)\cdots(1-q^k)} \tag{4.7.6}$$

where for a given partition δ we let $|\delta|$ denote the sum of its parts. Tus if $k(\delta)$ denotes the number of parts of δ we may also write

$$\sum_{\delta \in \Delta_2} x^{2k(\delta)} q^{|\delta|} = \sum_{k \ge 0} \frac{x^k q^{k^2}}{(1-q)(1-q^2)\cdots(1-q^k)} \, . \tag{4.7.7}$$

We may then translate (4.4.17) into the following remarkable identity:

Theorem 4.12

$$J(x; q) = \cfrac{1}{1 - \cfrac{qx^2}{1 - \cfrac{q^2x^2}{1 - \cfrac{q^3x^2}{1 - \cdots}}}} = \frac{\displaystyle\sum_{k \ge 0} \frac{(-1)^k x^{2k} q^{k^2+k}}{(1-q)(1-q^2)\cdots(1-q^k)}}{\displaystyle\sum_{k \ge 0} \frac{(-1)^k x^{2k} q^{k^2}}{(1-q)(1-q^2)\cdots(1-q^k)}}$$

$$\tag{4.7.8}$$

Proof Since the coefficient of $(-1)^k x^{2k}$ in $D_n(x; q)$ is given by the sum

$$\sum_{\substack{\delta \in \Delta_2(k) \\ \max \delta \le n}} q^{|\delta|}$$

and the latter converges to the righthand side of (4.7.6) (in the formal power series sense), we see that we must have

$$\lim_{n \to \infty} D_n(x; q) = \sum_{k \ge 0} \frac{(-1)^k x^{2k} q^{k^2}}{(1-q)(1-q^2)\cdots(1-q^k)} \, . \tag{4.7.9}$$

Now, as we have seen, the nth convergent of $J(x; q)$ is given by the expression

$$J^{(n)}(x; q) = \frac{N_n(x; q)}{D_n(x; q)} \qquad (4.7.10)$$

with

$$N_n(x; q) = \sum_{F \in \mathcal{F}([1,n], M)} (-1)^{|F|} w(F) \Big|_{\substack{d_i = \lambda_i x^2 \\ c_i = 0}} .$$

Since the latter is the image of $D_{n-1}(x; q)$ by the shift operator S, we immediately deduce that, we can obtain $N_n(x; q)$ by replacing all the q^i contributing to $D_{n-1}(x; q)$ by q^{i+1} and this gives that the coefficient of $(-1)^k x^{2k}$ in $N_n(x; q)$ must be equal to

$$\sum_{\substack{\delta \in \Delta_2(k) \\ \max \delta \le n-1}} q^{|\delta| + k(\delta)} .$$

Thus we must have

$$\lim_{n \to \infty} N_n(x; q) = \sum_{k \ge 0} \frac{(-1)^k x^{2k} q^{k^2 + k}}{(1 - q)(1 - q^2) \cdots (1 - q^k)} . \qquad (4.7.11)$$

Now, (4.5.19) in this case becomes

$$J^{(n)}(x; q) - J^{(n-1)}(x; q) = \frac{q^{\binom{n+1}{2}} x^{2n}}{D_n(x; q) D_{n-1}(x; q)}$$

and this implies that $J^{(n)}(x; q) \to J(x; q)$ as a formal power series in x or q as we desire. Combining this with (4.7.9) and (4.7.11) yields (4.7.8) and completes our proof. □

Note that since this proof shows that we can interpret (4.7.8) as a limit theorem involving formal power series in q we can set $x = i$ (the square root of -1) in (4.7.8) and derive the Rogers–Ramanujan continued fraction identity. More precisely we have

Theorem 4.13 *Let $D_n(q)$ and $N_n(q)$ denote the sequences defined by the recursions*

$$\begin{aligned} D_n(q) &= D_{n-1}(q) + q^n D_{n-2}(q) , \\ N_n(q) &= N_{n-1}(q) + q^n N_{n-2}(q) \end{aligned} \qquad \text{(for } n \ge 2) \qquad (4.7.12)$$

and the initial conditions

$$D_0(q) = 1, \; D_1(q) = 1 + q \; , \qquad N_0(q) = 1, \; N_1(q) = 1 \qquad (4.7.13)$$

Then

$$(a) \; \lim_{n \to \infty} D_n(q) = \sum_{k \geq 0} \frac{q^{k^2}}{(1 - q)(1 - q^2) \cdots (1 - q^k)} \; ,$$

$$(b) \; \lim_{n \to \infty} N_n(q) = \sum_{k \geq 0} \frac{q^{k^2 + k}}{(1 - q)(1 - q^2) \cdots (1 - q^k)} \; .$$

In particular we must have

$$\cfrac{1}{1 + \cfrac{q}{1 + \cfrac{q^2}{1 + \cfrac{q^3}{1 + \cdots}}}} = \frac{\displaystyle\sum_{k \geq 0} \frac{q^{k^2 + k}}{(1 - q)(1 - q^2) \cdots (1 - q^k)}}{\displaystyle\sum_{k \geq 0} \frac{q^{k^2}}{(1 - q)(1 - q^2) \cdots (1 - q^k)}} \; . \qquad (4.7.14)$$

We should mention that Rogers and Ramanujan (see [1]) showed that

$$\prod_{n \geq 1} (1 - q^n) \sum_{k \geq 0} \frac{q^{k^2}}{(1 - q)(1 - q^2) \cdots (1 - q^k)} = \sum_{m = -\infty}^{+\infty} (-1)^m \, q^{(5m^2 - m)/2} \; . \qquad (4.7.15)$$

This combined with the Jacobi triple product identity

$$\sum_{m = -\infty}^{+\infty} (-1)^m \, q^{\binom{n}{2}} z^m = \prod_{n \geq 0} (1 - q^{n+1})(1 - zq^n)(1 - q^{n+1}/z) \qquad (4.7.16)$$

with q replaced by q^5 and $z = q^2$ yields

$$\prod_{n \geq 1} (1 - q^n) \sum_{k \geq 0} \frac{q^{k^2}}{(1 - q)(1 - q^2) \cdots (1 - q^k)}$$

$$= \sum_{m = -\infty}^{+\infty} (-1)^m \, q^{(5m^2 - m)/2}$$

$$= \prod_{n \geq 0} (1 - q^{5n+5})(1 - q^{5n+2})(1 - q^{5n+3}) \; .$$

Now this, upon division by $\prod_{n\geq 1}(1-q^n)$, yields

$$\sum_{k\geq 0}\frac{q^{k^2}}{(1-q)(1-q^2)\cdots(1-q^k)}=\prod_{n\geq 0}\frac{1}{(1-q^{5n+1})(1-q^{5n+4})} \qquad (4.7.17)$$

which is usually referred to as the first Rogers–Ramanujan identity. The identity is actually due to Rogers who was the first to formulate it and *prove* it in 1894, about 20 years before Ramanujan formulated it without proof. The misnomer appears to be due to Hardy. From the combinatorial point of view the best proof of (4.7.15) was given by Schur who in 1917 rediscovered (4.7.17) and proved it in two different ways [21]. So historically speaking, this identity should be referred to as the *Rogers–Ramanujan–Schur identity* or perhaps even as the *Rogers–Schur–Ramanujan identity*.

There are several combinatorial proofs of the Jacobi identity. One of them can be found in [11], Schur proof of (4.7.15) can also be found in [11].

Now in the same manner it can be shown that

$$\sum_{k\geq 0}\frac{q^{k^2+k}}{(1-q)(1-q^2)\cdots(1-q^k)}=\prod_{n\geq 0}\frac{1}{(1-q^{5n+2})(1-q^{5n+3})}. \qquad (4.7.18)$$

Using (4.7.17) and (4.7.18) the identity in (4.7.14) can be rewritten in the truly remarkable form

$$\cfrac{1}{1+\cfrac{q}{1+\cfrac{q^2}{1+\cfrac{q^3}{1+\cdots}}}}=\prod_{n\geq 0}\frac{(1-q^{5n+1})(1-q^{5n+4})}{(1-q^{5n+2})(1-q^{5n+3})}. \qquad (4.7.19)$$

We should mention that a direct proof of (4.7.18) which avoids (4.7.15) but not the triple product identity can be found in [2].

4.8 Partitions and Hermite Polynomials

Our goal here is to present a beautiful bijection due to Flajolet [8] between set partitions and labeled Motzkin paths and as a biproduct obtain the continued fraction corresponding to the Hermite polynomials.

To this end we need some auxiliary constructions and notations. We shall begin with representing a subset $S=\{i_1<i_2<\ldots<i_s\}\subseteq\{1,2,\ldots,n\}$ as the line diagram obtained by placing s successive tiny circles $c_1, c_2, \ldots, c_s$ joined by line segments, with the center of circle c_r at a point of x-coordinate i_r. For instance, this

construction applied to the set $S = \{1, 3, 6, 7, 8\}$ gives the diagram in Fig. (4.8.1).

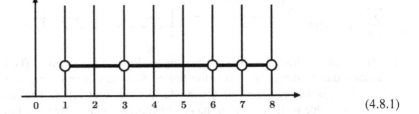

(4.8.1)

It will be convenient to refer to min S and max S (1 and 8 here) as the *opening* and *closing* elements of the diagram and refer to the others as *inner* elements.

It is customary to write the parts of a set partition in order of increasing minimal elements, we shall follow this convention here. This given, we can represent a set partition $\mathcal{A} = \{A_1, A_2, \ldots, A_m\}$ of $\{1, 2, \ldots, n\}$ by a juxtaposition of line diagrams with the kth diagram representing part A_k and drawn on the line $y = k$. The resulting composite diagram will be denoted by $\mathcal{D}(\mathcal{A})$. For instance, in this manner, the partition

$$\mathcal{A} = \big\{\{1, 2, 5, 9\}, \{3, 4\}, \{6\}, \{7, 10, 12, 13\}, \{8, 11, 14\}\big\}$$

is represented by Fig. (4.8.2).

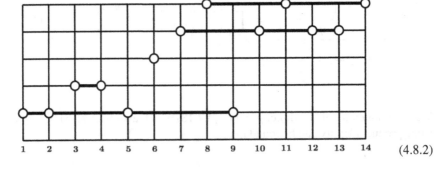

(4.8.2)

We see that if an element i of $\{1, 2, \ldots, n\}$ belongs to part A_h then the circle whose x-coordinate is i will lie at height h. We shall refer to the latter as the *height* of i and denote it by $h(i)$. We shall also say that an element i of $\{1, 2, \ldots, n\}$ is *covered* by a part A_r if and only if min $A_r \le i \le$ max A_r. Note that although the partition above has a total of 5 parts, the element 11 is covered only by two parts. It will be convenient to order the parts covering a given element i according to their heights in the diagram. For instance, the second part covering 8 in the picture above is $\{7, 10, 12, 13\}$.

The next step in Flajolet's construction is to convert $\mathcal{D}(\mathcal{A})$ into a *labeled* Motzkin path $L\pi(\mathcal{A})$. The corresponding unlabeled path $\pi(\mathcal{A})$ is easiest to describe in terms of its Motzkin word $w(\mathcal{D}(\mathcal{A}))$. We simply let

$$w(\mathcal{D}(\mathcal{A})) = d_1 d_2 \cdots d_n \qquad (4.8.3)$$

where, given that i is covered by k parts of $\mathcal{D}(\mathcal{A})$, we set

$$d_i = \begin{cases} a_k & \text{if } i \text{ is an } opening \text{ element,} \\ b_k & \text{if } i \text{ is a } closing \text{ element,} \\ c_k & \text{if } i \text{ is an } inner \text{ element,} \\ c_k & \text{if } i \text{ is a } singleton \text{ part of } \mathcal{A}. \end{cases} \qquad (4.8.4)$$

Carrying this out for the partition in (4.8.2) yields the word

$$a_0 c_1 a_1 b_2 c_1 c_1 a_1 a_2 b_3 c_2 c_2 c_2 b_2 b_1 \ .$$

Recalling that the letters a_k, b_k, c_k represent respectively NORTHEAST, SOUTH-EAST and EAST steps of the path, we see that the partition diagram in (4.8.2) can thus be converted into the following Motzkin path

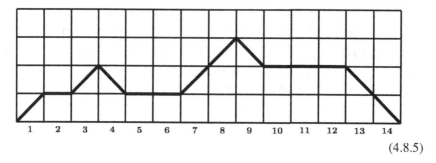

$$(4.8.5)$$

The final object of this construction, that is the labeled Motzkin path $L\pi(\mathcal{A})$ is simply obtained by labeling the edges of $\pi(\mathcal{A})$ according to the following rules. If e_i is the ith edge of $\pi(\mathcal{A})$ then the label of e_i is

$$L(e_i) = \begin{cases} z_0 & \text{if } i \text{ is an } opening \text{ element,} \\ k & \text{if } i \text{ } closes \text{ the } k^{th} \text{ part covering } i, \\ k & \text{if } i \text{ is an } inner \text{ of the } k^{th} \text{ part covering } i, \\ z_s & \text{if } i \text{ is a } singleton \text{ part of } \mathcal{A}. \end{cases} \qquad (4.8.6)$$

This procedure converts the path in (4.8.5) into the labeled path given below.

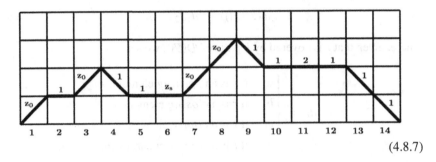

(4.8.7)

It is not difficult to see that this construction yields a bijection between partitions of the set $\{1, 2, \ldots, n\}$ and labeled Motzkin paths in $\Pi_{0,0}(n)$ where the label $L(e)$ of an edge starting at height k is

$$L(e) = \begin{cases} z_0 & \text{if } e \text{ is a NORTHEAST edge,} \\ \in \{1, 2, \ldots, k\} & \text{if } e \text{ is a SOUTHEAST edge,} \\ \in \{z_s, 1, 2, \ldots, k\} & \text{if } e \text{ is an EAST edge.} \end{cases} \qquad (4.8.8)$$

This observation can be translated into the following remarkable result

Theorem 4.14 *(Flajolet) If μ_{n,n_0,n_i,n_s} denotes the number of partitions of the set $\{1, 2, \ldots, n\}$ with n_0 nonsingleton parts, n_i inner elements and n_s singletons then we have the following continued fraction expansion*

$$\cfrac{1}{1 - z_s x - \cfrac{1 \cdot z_0 x^2}{1 - (z_s + z_i)x - \cfrac{2 \cdot z_0 x^2}{1 - (z_s + 2z_i)x - \cfrac{3 \cdot z_0 x^2}{1 - (z_s + 3z_i)x - \cdots}}}}$$

$$= \sum_{n=2n_0+n_i+n_s} \mu_{n,n_0,n_i,n_s} \, x^n z_0^{n_0} z_i^{n_i} z_s^{n_s} \, . \qquad (4.8.9)$$

Proof Denoting by $\mathcal{P}_n$ the collection of all partitions of $\{1, 2, \ldots, n\}$, we see that the right hand side of (4.8.9) is none other than

$$\sum_{n \geq 0} x^n \sum_{A \in \mathcal{P}_n} z_0^{n_0(A)} z_i^{n_i(A)} x_s^{n_s(A)} \qquad (4.8.10)$$

where $n_0(A)$, $n_i(A)$ and $n_s(A)$ respectively denote the number of nonsingleton parts, the number of inner elements and the number of singleton parts of A. The Flajolet correspondence allows us to rewrite (4.8.10) as a sum over labeled Motzkin parts. That is

$$\text{RHS of (4.8.9)} = \sum_{n \geq 0} x^n \sum_{L\pi \in L\Pi_{0,0}(n)} p(L\pi) , \qquad (4.8.11)$$

where $L\Pi_{0,0}(n)$ denotes the collection of all labeled paths obtained by taking each path in $\Pi_{0,0}(n)$ and labeling it in all possible ways consistent with the requirements in (4.8.8). Here of course we must define $p(L\pi)$ so that if $L\pi$ is the labeled path corresponding to a partition A then

$$p(L\pi) = z_0^{n_0(A)} z_i^{n_i(A)} x_s^{n_s(A)} .$$

Let $L\pi$ denote the generic labeled path obtained by labeling π according to (4.8.8) and set

$$P(\pi) = \sum_{L\pi} p(L\pi) .$$

It is not difficult to see that the contribution to $P(\pi)$ of an edge e of π, which starts at level k, is a factor $f(e)$ which is given by

$$f(e) = \begin{cases} z_0 & \text{if } e \text{ is a NORTHEAST edge,} \\ k & \text{if } e \text{ is a SOUTHEAST edge,} \\ z_s + kz_i & \text{if } e \text{ is an EAST edge.} \end{cases}$$

This yields that

$$\sum_{L\pi \in L\Pi_{0,0}(n)} p(L\pi) = \sum_{\pi \in \Pi_{0,0}(n)} w(\pi) \Bigg|_{\substack{a_k=z_0 \\ b_k=k \\ c_k=z_s+kz_i}} .$$

In other words, the right hand side of (4.8.9) is simply equal to the formal series obtained by setting $a_k = z_0, b_k = k, c_k = z_s + kz_i$ in the right hand side of (4.5.2). But then Eq. (4.5.4) gives that we can also obtain the same result by making these replacements in the continued fraction

$$\cfrac{1}{1 - c_0 x - \cfrac{a_0 b_1 x^2}{1 - c_1 x - \cfrac{a_1 b_2 x^2}{1 - c_2 x - \cfrac{a_2 b_3 x^2}{1 - c_3 x - \cdots}}}} .$$

However doing this gives precisely the right hand side of (4.8.9) as desired. $\qquad \square$

Theorem 4.14 yields the continued fraction expansion for the generating functions of the Bell numbers B_n, the Stirling numbers $S_{n,k}$ and the numbers I_n and J_n of involutions of S_n with and without fixed points. These are the contents of the four theorems given below.

Theorem 4.15 *Denoting by B_n the number of partitions of $\{1, 2, \ldots, n\}$ we have*

$$\cfrac{1}{1 - x - \cfrac{1 \cdot x^2}{1 - 2 \cdot x - \cfrac{2 \cdot x^2}{1 - 3 \cdot x - \cfrac{3 \cdot x^2}{1 - 4 \cdot x - \cdots}}}}$$

$$= \sum_{n \geq 0} B_n x^n \ . \tag{4.8.12}$$

Proof Just set $z_0 = z_i = z_s = 1$ in (4.8.9). $\qquad\square$

Theorem 4.16 *If $S_{n,k}$ denotes the number of partitions of $\{1, 2, \ldots, n\}$ into k parts, then*

$$\cfrac{1}{1 - tx - \cfrac{1 \cdot tx^2}{1 - (t+1)x - \cfrac{2 \cdot tx^2}{1 - (t+2)x - \cfrac{3 \cdot tx^2}{1 - (t+3)x - \cdots}}}} = \sum_{n \geq 0} x^n \sum_{k=1}^{n} S_{n,k} t^k \ .$$

$$\tag{4.8.13}$$

Proof Just set $z_i = 1$ and $z_0 = z_s = t$ in (4.8.9). $\qquad\square$

Theorem 4.17 *If I_n denotes the number of involutions in the symmetric group S_n then*

$$\cfrac{1}{1 - x - \cfrac{1 \cdot x^2}{1 - x - \cfrac{2 \cdot x^2}{1 - x - \cfrac{3 \cdot x^2}{1 - x - \cdots}}}} = \sum_{n \geq 0} I_n x^n \ . \tag{4.8.14}$$

Proof Note that if we set $z_i = 0$ and $z_0 = z_s = 1$ in (4.8.9) then the right hand side of (4.8.9) yields the generating function of the number of partitions of $\{1, 2, \ldots, n\}$ with no parts of cardinality greater than two. But the latter partitions are clearly in bijection with involutions. $\qquad\square$

Theorem 4.18 *If J_n denotes the number of involutions in the symmetric group S_n which have no fixed points then $J_n = 0$ for n odd and*

$$\cfrac{1}{1 - \cfrac{1 \cdot x^2}{1 - \cfrac{2 \cdot x^2}{1 - \cfrac{3 \cdot x^2}{1 - \cdots}}}} = \sum_{n \geq 0} J_{2n} x^{2n} \; . \tag{4.8.15}$$

Proof By the same reasons we gave in the previous proof we can obtain this result by setting $z_s = z_i = 0$ and $z_0 = 1$ in (4.8.9). □

Remark 4.2 It is well known that the number of permutations of S_n with α_i cycles of length i is given by the expression

$$\frac{n!}{\prod_i i^{\alpha_i} \alpha_i!} \; .$$

This gives that the number of involutions of S_n with k 2-cycles is

$$\frac{n!}{(n - 2k)! \, 2^k k!} \; .$$

Thus setting $z_s = 1$, $z_i = 0$ and $z_0 = t$ in (4.8.9) gives the identity

$$\cfrac{1}{1 - x - \cfrac{1 \cdot tx^2}{1 - x - \cfrac{2 \cdot tx^2}{1 - x - \cfrac{3 \cdot tx^2}{1 - x - \cdots}}}} = \sum_{n \geq 0} x^n \sum_{k \leq n/2} \frac{n!}{(n - 2k)! \, 2^k k!} t^k \; .$$

$$\tag{4.8.16}$$

Similarly we get that the number J_{2n} of involutions without fixed points in S_{2n} is

$$\frac{(2n)!}{2^n n!} = 1 \cdot 3 \cdot 5 \cdots (2n - 1) \; .$$

Thus formula (4.8.15) may be read as

$$\cfrac{1}{1 - \cfrac{1 \cdot x^2}{1 - \cfrac{2 \cdot x^2}{1 - \cfrac{3 \cdot x^2}{1 - \cdots}}}} = \sum_{n \geq 0} 1 \cdot 3 \cdot 5 \cdots (2n - 1) \, x^{2n} \; . \tag{4.8.17}$$

This last observation yields us the connection between involutions and the Hermite polynomials $H_n(x)$. The latter may be defined by letting

$$H_n(x) = (-1)^n e^{x^2/2} D^n e^{-x^2/2} . \qquad (4.8.18)$$

Note that a single integration by parts gives that

$$\frac{1}{\sqrt{\pi}} \int_{-\infty}^{+\infty} x^m H_n(x) e^{-x^2/2} dx = \frac{m}{\sqrt{\pi}} \int_{-\infty}^{+\infty} x^{m-1} H_{n-1}(x) e^{-x^2/2} dx$$

If $m \le n$ we can apply this relation m times in succession and obtain that

$$\frac{1}{\sqrt{\pi}} \int_{-\infty}^{+\infty} x^m H_n(x) e^{-x^2/2} dx = \begin{cases} n! & \text{if } m = n, \\ 0 & \text{if } m < n . \end{cases} \qquad (4.8.19)$$

This implies that the Hermite polynomials are orthogonal with respect to the measure μ on $(-\infty, +\infty)$ given by setting

$$d\mu = \frac{1}{\sqrt{\pi}} e^{-x^2/2} dx . \qquad (4.8.20)$$

Since the leading term in $H_n(x)$ is x^n we can apply the theory directly here and obtain

Theorem 4.19 *Let $H_n^*(x) = x^n H_n(\frac{1}{x})$ and let $K_n^*(x)$ be the sequence of polynomials defined by the recursion $K_{n+1}^*(x) = K_n^*(x) - nx^2 K_{n-1}^*(x)$ with initial conditions $K_0^*(x) = K_1^*(x) = 1$ then*

$$\lim_{n \to \infty} \frac{K_n^*(x)}{H_{n+1}^*(x)} = \cfrac{1}{1 - \cfrac{1 \cdot x^2}{1 - \cfrac{2 \cdot x^2}{1 - \cfrac{3 \cdot x^2}{1 - \cdots}}}} \qquad (4.8.21)$$

In other words this is the continued fraction corresponding to the Hermite polynomials. In particular, they must satisfy the recursion

$$H_{n+1}(x) = x H_n(x) - n H_{n-1}(x) . \qquad (4.8.22)$$

Proof From the continued fraction expansion in (4.8.17) and Theorem 4.5) it follows that the sequence of polynomials $\{Q_n(x)\}$ orthogonal with respect to a measure with moments μ_n given by

$$\begin{cases} (a) \ \mu_{2n-1} = 0 \,, \\ (b) \ \mu_{2n} = 1 \cdot 3 \cdots (2n-1) \end{cases} \tag{4.8.23}$$

and satisfying the requirement in (4.4.3) must satisfy the recurrence $Q_{n+1} = xQ_n - nQ_{n-1}$. So to prove our assertions we need only verify the measure μ given in (4.8.20) has precisely these moments. Now (4.8.23) (a) is trivial since μ is symmetrically distributed around the origin. As for (4.8.23) (b) we only need to observe that it follows from the recursion

$$\frac{1}{\sqrt{\pi}} \int_{-\infty}^{+\infty} x^{2n} e^{-x^2/2} \, dx = \frac{2n-1}{\sqrt{\pi}} \int_{-\infty}^{+\infty} x^{2n-2} e^{-x^2/2} \, dx \,,$$

(which we can immediately obtain by a single integration by parts) and the initial condition $\mu_0 = 1$ which is given by the classical integral

$$\int_{-\infty}^{+\infty} e^{-x^2/2} \, dx = \sqrt{\pi} \,. \qquad \qquad \square$$

This completes our section.

4.9   The Legendre Polynomials

The Legendre polynomials $\{L_n(x)\}$ may defined by means of the formula

$$L_n(x) = \frac{1}{2^n n!} D^n (x^2 - 1)^n \,. \tag{4.9.1}$$

These polynomials are easily shown to be orthogonal with respect to the measure $d\mu = dx/2$ concentrated on the interval $[-1, 1]$. Indeed, we do have that

Proposition 4.9

$$\frac{1}{2} \int_{-1}^{1} P_n(x) P_m(x) \, dx = \begin{cases} 0 & \text{if } n \neq m, \\ \\ \frac{1}{2n+1} & \text{if } n = m \,. \end{cases} \tag{4.9.2}$$

Proof For convenience let $\Gamma_n = D^n (x^2 - 1)^n$. Then an integration by parts gives

$$\int_{-1}^{1} \Gamma(x) \, \Gamma_m(x) \, dx = \int_{-1}^{1} D^{n-1}(x^2 - 1)^n \, D^{m+1}(x^2 - 1)^m \, dx$$

and repeating this process will necessarily yield zero after $m+1$ integration by parts if $m < n$. This establishes the first equality in (4.9.2). On the other hand if $m = n$ then n successive integrations by parts yield

$$\int_{-1}^{1} \Gamma_n(x) \, \Gamma_n(x) \, dx = (-1)^n \, (2n)! \int_{-1}^{1} (x^2 - 1)^n \, dx \ . \tag{4.9.3}$$

But integrating by parts again we get

$$\int_{-1}^{1} (1 - x^2)^n \, dx = 2n \int_{-1}^{1} (1 - x^2)^{n-1} x^2 \, dx \ .$$

Now this can be rewritten in the form

$$(2n + 1) \int_{-1}^{1} (1 - x^2)^n \, dx = 2n \int_{-1}^{1} (1 - x^2)^{n-1} \, dx \ .$$

Iterating and combining with (4.9.3) yields

$$\int_{-1}^{1} \Gamma_n^2(x) \, dx = 2 \frac{(2n)!!}{(2n + 1)!!} \, (2n)! = 2 \frac{((2n)!!)^2}{2n + 1} \ .$$

Since $\Gamma_n = 2^n \, n! \, P_n$ we see that this is precisely the second identity in (4.9.2). $\quad\square$

Now the definition in (4.9.1) gives that the leading term in $L_n(x)$ is $\frac{(2n-1)!!}{n!} x^n$. Thus to apply our machinery we need to take

$$Q_n(x) = \frac{n!}{(2n - 1)!!} L_n(x) \ . \tag{4.9.4}$$

We can then rewrite (4.9.2) (2) in terms of the Q_n to read

$$\int_{-1}^{1} Q_n^2(x) \, dx = \frac{(n!)^2}{(2n - 1)!! \, (2n + 1)!!} \ . \tag{4.9.5}$$

We are now in a position to determine the coefficients c_n and λ_n occurring in the three-term recursion. Indeed, we need not work very hard for c_n since P_n and therefore also Q_n is an odd function of x for odd n and an even function for n even. Thus, since μ is symmetrically distributed, the integral formula (4.4.10) gives that

$c_n = 0$ for all n. As for λ_n, we can apply formula (4.5.32) (a) and obtain from (4.9.5) that

$$\lambda_n = \frac{\int_{-1}^{1} Q_n^2(x)\,dx}{\int_{-1}^{1} Q_{n-1}^2(x)\,dx} = \frac{n^2}{(2n-1)(2n+1)} \, . \tag{4.9.6}$$

This gives the recursion

$$Q_{n+1} = x\, Q_n - \frac{n^2}{(2n-1)(2n+1)}\, Q_{n-1} \, .$$

Using (4.9.4) we deduce that the Legendre polynomials themselves must satisfy the recursion

$$(n+1)\, L_n(x) = (2n+1)x\, L_n(x) - n\, L_{n-1}(x) \, . \tag{4.9.7}$$

Note further that in this case the sequence of moments is given by

$$\mu_n = \frac{1}{2} \int_{-1}^{1} x^n\, dx = \begin{cases} 0 & \text{if } n \text{ is odd,} \\ \frac{1}{(n+1)} & \text{if } n \text{ is even.} \end{cases}$$

Thus this time Theorem 4.5 may be used to derive a rather neat continued fraction expansion. Namely

Theorem 4.20

$$\cfrac{x}{1 - \cfrac{1^2 \cdot x^2/1 \cdot 3}{1 - \cfrac{2^2 \cdot x^2/3 \cdot 5}{1 - \cfrac{3^2 \cdot x^2/5 \cdot 7}{1 - \cdots}}}} = \sum_{k\geq 0} \frac{x^{2k+1}}{2k+1} = \log \sqrt{\frac{1+x}{1-x}} \, . \tag{4.9.8}$$

4.10   The Laguerre Polynomials

These polynomials depend on a parameter α which may be assumed to be an independent variable. They can be defined by setting

$$L_n^{(\alpha)}(x) = (-1)^n\, e^x\, x^{-\alpha}\, D^n \left(x^{n+\alpha}\, e^{-x}\right) \, . \tag{4.10.1}$$

Note that an integration by parts, gives

$$\int_0^{+\infty} x^m \, L_n^{(\alpha)}(x) x^\alpha \, e^{-x} \, dx = (-1)^{n-1} \, m \int_0^{+\infty} x^{m-1} \, D^{n-1}\left(x^{n+\alpha} \, e^{-x}\right) dx \ .$$

(4.10.2)

We see again that if $m < n$ then $m + 1$ successive integration by parts yield that this integral vanishes. If $m \geq n$ we can carry out n successive integrations by parts and get

$$\int_0^{+\infty} x^m \, L_n^{(\alpha)}(x) x^\alpha \, e^{-x} \, dx = \frac{m!}{(m-n)!} \int_0^{+\infty} x^{m+\alpha} \, e^{-x} \, dx \ .$$

Finally, m more integrations by parts give

$$\int_0^{+\infty} x^m \, L_n^{(\alpha)}(x) \, x^\alpha e^{-x} \, dx \tag{4.10.3}$$

$$= \frac{m!}{(m-n)!} \, (m+\alpha)(m+\alpha-1)\cdots(\alpha+1) \int_0^{+\infty} x^\alpha \, e^{-x} \, dx \ .$$

Letting μ be the measure concentrated on $[0, \infty)$ defined by setting

$$d\mu = \frac{x^\alpha \, e^{-x}}{\Gamma(\alpha+1)} \quad \text{with} \quad \Gamma(\alpha+1) = \int_0^{+\infty} x^\alpha \, e^{-x} \, dx \ ,$$

we can translate our findings here into the identities

$$\int_0^{+\infty} x^m L_n^{(\alpha)}(x) d\mu = \begin{cases} 0 & \text{if } m < n, \\ \frac{m!}{(m-n)!}(m+\alpha)(m+\alpha-1)\cdots(\alpha+1) & \text{if } m \geq n . \end{cases}$$

(4.10.4)

In particular, we see that the polynomials $L_n^{(\alpha)}(x)$ are orthogonal with respect to μ. Moreover, since the leading term in $L_n^{(\alpha)}(x)$ is again x^n we can apply our theory without rescaling. Now formulas (4.5.34) and (4.10.4) for $n = m$ immediately give that

$$\lambda_n = \frac{n! \, (n+\alpha)(n+\alpha-1)\cdots(\alpha+1)}{(n-1)! \, (n+\alpha-1)(n+\alpha-1)\cdots(\alpha+1)} = n(n+\alpha) \ . \tag{4.10.5}$$

To compute c_n here we shall use (4.5.32) which we rewrite in the form

$$c_n = \frac{\chi_n}{d_{n-1}} \frac{d_{n-1}}{d_n} - \frac{\chi_{n-1}}{d_{n-2}} \frac{d_{n-2}}{d_{n-1}} \ . \tag{4.10.6}$$

Now (4.5.37) gives that

$$\frac{\chi_n}{d_{n-1}} = \frac{(-1)^n}{\Gamma(\alpha+1)} \int_0^\infty x^{n+1} D^n x^{\alpha+n} e^{-x} \, dx \ ,$$

and using (4.10.4) for $m = n + 1$ we get

$$\frac{\chi_n}{d_{n-1}} = (n+1)! \, (\alpha+n+1)(\alpha+n) \cdots (\alpha+1) \ . \tag{4.10.7}$$

On the other hand (4.5.35) and (4.10.4) for $m = n$ give

$$\frac{d_n}{d_{n-1}} = n! \, (\alpha+n)(\alpha+n-1) \cdots (\alpha+1) \ . \tag{4.10.8}$$

Using (4.10.7) and (4.10.8) for n and $n-1$ we can finally conclude that

$$
\begin{aligned}
c_n &= \frac{(n+1)! \, (\alpha+n+1)(\alpha+n) \cdots (\alpha+1)}{n! \, (\alpha+n)(\alpha+n-1) \cdots (\alpha+1)} \\
&\quad - \frac{n! \, (\alpha+n)(\alpha+n-1) \cdots (\alpha+1)}{(n-1)! \, (\alpha+n-1)(\alpha+n-2) \cdots (\alpha+1)} \\
&= 2n + \alpha + 1 \ .
\end{aligned} \tag{4.10.9}
$$

The moments μ_n are also easily computed in this case. In fact, we have

$$\mu_n = \frac{1}{\Gamma(\alpha+1)} \int_0^{+\infty} x^{n+\alpha} e^{-x} \, dx = (\alpha+n)(\alpha+n-1) \cdots (\alpha+1) \ . \tag{4.10.10}$$

Formulas (4.10.5), (4.10.6), and (4.10.10) yield us the continued fraction expansion in Theorem 4.21.

Theorem 4.21

$$
\cfrac{1}{1 - (\alpha+1)x - \cfrac{1 \cdot (\alpha+1)x^2}{1 - (\alpha+3)x - \cfrac{2 \cdot (\alpha+2)x^2}{1 - (\alpha+5)x - \cfrac{3 \cdot (\alpha+3)x^2}{1 - (\alpha+7)x - \cdots}}}}
$$

$$= \sum_{n \geq 0} (\alpha+n)(\alpha+n-1) \cdots (\alpha+1) \, x^n \ . \tag{4.10.11}$$

Note that letting $\alpha = 0$ in (4.10.11) yields the remarkable identity

$$\cfrac{1}{1 - 1 \cdot x - \cfrac{1^2 \cdot x^2}{1 - 3 \cdot x - \cfrac{2^2 \cdot x^2}{1 - 5 \cdot x - \cfrac{3^3 \cdot x^2}{1 - 7 \cdot x - \cdots}}}} = \sum_{n \geq 0} n! \, x^n \, . \qquad (4.10.12)$$

In closing we should note that some authors, whose name may be best left out here, have had some difficulty swallowing this identity. Their problem was that they could not see how to fit it into their Cauchy-limited world. Apparently Gauss had no such difficulty for he knew very well how to make sense out of it.

From our point of view (4.10.12) in none other than a purely combinatorial identity. It simply expresses the existence of a bijection between permutations and a certain class of labeled Motzkin paths. In fact, using the interpretation of the moments given in (4.5.5), we can see that (4.10.11) can be established by showing that

$$\sum_{\pi \in \Pi_{0,0}(n)} w(\pi) \, \Bigg|_{\substack{a_n = n+1 \\ b_n = n+\alpha \\ c_n = 2n+1+\alpha}} = (\alpha + n)(\alpha + n - 1) \cdots (\alpha + 1) \, . \qquad (4.10.13)$$

To get this we need a bijection between permutations in S_n and labeled Motzkin paths in $\Pi_{0,0}(n)$ where the label $L(e)$ of an edge e which starts at height k should be

$$L(e) = \begin{cases} \in \{1, 2, \ldots, k\} & \text{if } e \text{ is a NORTHEAST edge,} \\ \in \{1, 2, \ldots, k, \alpha\} & \text{if } e \text{ is a SOUTHEAST edge,} \\ \in \{1, 2, \ldots, 2k + 1, \alpha\} & \text{if } e \text{ is an EAST edge .} \end{cases} \qquad (4.10.14)$$

Now a very beautiful bijection doing precisely that was given by Françon and Viennot in [9]. It can also be found in Flajolet's paper [8].

We should point out that, just as was the case for the Flajolet bijection given in Sect. 4.8, The Françon–Viennot bijection yields a variety of continued fraction expansions. Using it we may obtain generating functions of combinatorial sequences enumerating permutations according to various statistics. Notably, it gives a combinatorial explanation to the Stieltjes computation of the Laplace transform of $\tan x$. However, in order to keep these notes to a reasonable size we shall have to refer the reader to the original papers for this additional material.

Lecture 5
Basics on Finite Fields

5.1 Introduction

This final set of lectures on finite fields is not aligned exactly with the character of the first four, but we are grateful that the editors have decided to include it here because it is an interesting manifestation of the spirit of exposition of the volume.

Recall that $\mathcal{F}$ is a field if and only if

(i) $\mathcal{F}$ it has two operations "$+$" and "$\times$" and two elements "0" and "1" such that both $(\mathcal{F}, +)$ and $(\mathcal{F} - \{0\}, \times)$ are abelian groups,

(ii) For all $x, y, z \in \mathcal{F}$ we have

$$x(y + z) = (y + z)x = xy + xz = yx + zx .$$

Note that the finiteness of $\mathcal{F}$ implies that the infinite sequence

$$1, \quad 1+1, \quad 1+1+1, \quad 1+1+1+1, \quad \dots$$

must contain only a finite number of distinct elements. Denoting by k the kth element of this sequence, then for some $h < k$ we must have $h = k$. This forces $k - h = 0$ in $\mathcal{F}$. The integer

$$p = \min\{k \mid k = 0 \text{ in } \mathcal{F}\}$$

is called the *characteristic of* $\mathcal{F}$. It is easily seen that

(iii) $k = 0$ in $\mathcal{F}$ if and only if p divides k,

(iv) p is a prime number.

© The Editor(s) (if applicable) and The Author(s), under exclusive license
to Springer Nature Switzerland AG 2020
A. M. Garsia, Ö. Eğecioğlu, *Lectures in Algebraic Combinatorics*,
Lecture Notes in Mathematics 2277, https://doi.org/10.1007/978-3-030-58373-6_5

In fact, note that if $k = 0$ in $\mathcal{F}$ then the definition of p forces $k \geq p$ and if $k > p$ then we can find two integers Q and R such that $k = Qp + R$ with $R < p$ yielding that $R = 0$ in $\mathcal{F}$. But then $R > 0$ would contradict the definition of p, proving (iii). Likewise if $p = ab$ with $a, b > 1$ then $a \neq 0$ in $\mathcal{F}$ and the fact that $(\mathcal{F} - \{0\}, \times)$ is a group would give

$$a^{-1} p = b \qquad (\text{in } \mathcal{F})$$

forcing $b = 0$ in $\mathcal{F}$, again contradicting the definition of p. Clearly we are led to the same conclusion if $b \neq 0$ in $\mathcal{F}$. This proves (iv).

Throughout the rest of this section $\mathcal{F}$ will denote a finite field and q its cardinality.

Before we can proceed with deeper properties of finite fields we must acquire some basic tools for working with polynomials. These will be presented in separate subsections.

5.2   The Euclidean Algorithm

The original Euclidean algorithm was a process which yielded the greatest common divisor d of two given integers A and B. We shall use it here for the ring of integers as well as for the ring of polynomials in a single variable x with coefficients in a field. We shall present here a modified version of the Euclidean algorithm due to Berlekamp [3] which is extremely elegant and very suitable for explicit computations of greatest common divisors. For the rest of this section all the elements we deal with are supposed to belong to some fixed ring $\mathcal{R}$ with $\mathcal{R} = \mathbb{Z}$ or $\mathcal{R} = \mathcal{F}[x]$ for some field $\mathcal{F}$. In order to use the same terminology in any of these cases, we let $\text{degree}(a) = \ln |a|$ for $a \in \mathbb{Z}$ and the ordinary degree of a polynomial when $a \in \mathcal{F}[x]$.

We recall, that the greatest common divisor of two elements $A, B \in \mathcal{R}$ is denoted $\gcd(A, B)$ and we say that A and B are *relatively prime* if and only if $\gcd(A, B) = 1$.

This given, to compute the greatest common divisor $\gcd(A, B)$ by the Berlekamp algorithm we start by setting

$$r_{-2} = A, \quad r_{-1} = B, \quad p_{-2} = 0, \quad p_{-1} = 1, \quad q_{-2} = 1, \quad q_{-1} = 0 \qquad (5.2.1)$$

and then compute a_k, r_k, p_k, q_k recursively as follows:

> (i) $r_{k-2} = a_k r_{k-1} + r_k$ (by the division algorithm)
>
> (ii) $p_k = a_k p_{k-1} + p_{k-2}$ $\qquad\qquad\qquad\qquad$ (5.2.2)
>
> (iii) $q_k = a_k q_{k-1} + q_{k-2}$

for $k = 0, 1, 2, \ldots$. Note that given two elements $a, b \in \mathcal{R}$, the *division* of a by b consists in the construction of two elements $q, r \in \mathcal{R}$, yielding $a = bq + r$

(usually referred to as the *quotient* and the *remainder*) and uniquely determined by the condition that $\text{degree}(r) < \text{degree}(b)$. This is what we mean by the division algorithm above.

Since the degree of r_k decreases at least by one at each step, after $n < B$ steps we shall have $r_n = 0$. we will show that these recursions force the following basic identitiesi:

$$
\begin{aligned}
(a) \ & q_n \, p_{n-1} - p_n \, q_{n-1} = (-1)^n \\
(b) \ & A = r_{n-1} \, p_n \\
(c) \ & B = r_{n-1} \, q_n \\
(d) \ & B \, p_{n-1} - A \, q_{n-1} = (-1)^n r_{n-1} \ .
\end{aligned}
\tag{5.2.3}
$$

Clearly we need only establish (a), (b) and (c) since (d) is obtained by multiplying the first of these equations by r_{n-1} and using $(5.2.3)\,b)$ and $(5.2.3)\,c$.

Now note that (b) and (c) show that r_{n-1} is a common divisor of A and B, while (d) shows that any common divisor of A and B must divide r_{n-1}. Thus

$$\gcd(A, B) = r_{n-1}$$

and we may rewrite $(5.2.3)$ (d) in the form

$$
\gcd(A, B) = H A - K B \qquad \text{with} \quad
\begin{cases}
H = (-1)^{n-1} q_{n-1} \\
K = (-1)^{n-1} p_{n-1}
\end{cases}
\tag{5.2.4}
$$

We should note that the advantage of this process over the one that is usually described in most textbooks is that it yields, not only the greatest common divisor of A and B but at the same time it constructs a pair of elements p_{n-1}, q_{n-1} yielding the expansion in $(5.2.4)$ expressing $\gcd(A, B)$ as a linear combination of A and B. Moreover it does all this without the need of excessive storage of partial results. In fact, only six results need to be stored at any particular time, a constant that is independent of the choice of A and B.

We should also mention that we can use the Euclidean algorithm to construct the continued fraction expansion of any rational number. The basic identity here may be written as

$$
\frac{A}{B} = a_0 + \cfrac{1}{a_1 + \cfrac{1}{a_2 + \cfrac{\cdots}{\cdots \cfrac{}{a_{n-1} + \cfrac{1}{a_n}}}}}
$$

To prove the validity of this algorithm we need only establish the relations in (5.2.3). To this end it is convenient to rewrite the recursions in (5.2.2) in matrix form. For instance we may write (5.2.2) (i) as

$$\begin{bmatrix} a_k & 1 \\ 1 & 0 \end{bmatrix} \times \begin{bmatrix} r_{k-1} \\ r_k \end{bmatrix} = \begin{bmatrix} r_{k-2} \\ r_{k-1} \end{bmatrix}. \tag{5.2.5}$$

Likewise (5.2.2) (ii) and (iii) can be simultaneously obtained from the single equation

$$\begin{bmatrix} p_k & p_{k-1} \\ q_k & q_{k-1} \end{bmatrix} = \begin{bmatrix} p_{k-1} & p_{k-2} \\ q_{k-1} & q_{k-2} \end{bmatrix} \times \begin{bmatrix} a_k & 1 \\ 1 & 0 \end{bmatrix}. \tag{5.2.6}$$

Setting $k = 0$ in (5.2.5) and using the initial conditions in (5.2.1) gives

$$\begin{bmatrix} a_0 & 1 \\ 1 & 0 \end{bmatrix} \times \begin{bmatrix} B \\ r_0 \end{bmatrix} = \begin{bmatrix} A \\ B \end{bmatrix}. \tag{5.2.7}$$

Similarly, setting $k = 1$ in (5.2.5) we get

$$\begin{bmatrix} a_1 & 1 \\ 1 & 0 \end{bmatrix} \times \begin{bmatrix} r_0 \\ r_1 \end{bmatrix} = \begin{bmatrix} B \\ r_0 \end{bmatrix}.$$

Now this may be used in (5.2.7) to give

$$\begin{bmatrix} a_0 & 1 \\ 1 & 0 \end{bmatrix} \times \begin{bmatrix} a_1 & 1 \\ 1 & 0 \end{bmatrix} \times \begin{bmatrix} r_0 \\ r_1 \end{bmatrix} = \begin{bmatrix} A \\ B \end{bmatrix}. \tag{5.2.8}$$

Next, setting $k = 2$ in (5.2.5) we get

$$\begin{bmatrix} a_2 & 1 \\ 1 & 0 \end{bmatrix} \times \begin{bmatrix} r_1 \\ r_2 \end{bmatrix} = \begin{bmatrix} r_0 \\ r_1 \end{bmatrix}$$

and using this in (5.2.8) gives

$$\begin{bmatrix} a_0 & 1 \\ 1 & 0 \end{bmatrix} \times \begin{bmatrix} a_1 & 1 \\ 1 & 0 \end{bmatrix} \times \begin{bmatrix} a_2 & 1 \\ 1 & 0 \end{bmatrix} \times \begin{bmatrix} r_1 \\ r_2 \end{bmatrix} = \begin{bmatrix} A \\ B \end{bmatrix}.$$

This should make it clear that repeating this process k times leads to the identity

$$\begin{bmatrix} a_0 & 1 \\ 1 & 0 \end{bmatrix} \times \begin{bmatrix} a_1 & 1 \\ 1 & 0 \end{bmatrix} \times \cdots \times \begin{bmatrix} a_k & 1 \\ 1 & 0 \end{bmatrix} \times \begin{bmatrix} r_{k-1} \\ r_k \end{bmatrix} = \begin{bmatrix} A \\ B \end{bmatrix}. \tag{5.2.9}$$

Likewise, k repeated applications of the recurrence in (5.2.6) yields that

$$\begin{bmatrix} p_k & p_{k-1} \\ q_k & q_{k-1} \end{bmatrix} = \begin{bmatrix} p_{-1} & p_{-2} \\ q_{-1} & q_{-2} \end{bmatrix} \times \begin{bmatrix} a_0 & 1 \\ 1 & 0 \end{bmatrix} \times \begin{bmatrix} a_1 & 1 \\ 1 & 0 \end{bmatrix} \times \cdots \times \begin{bmatrix} a_k & 1 \\ 1 & 0 \end{bmatrix}. \qquad (5.2.10)$$

Now the initial conditions in (5.2.1) give

$$\begin{bmatrix} p_{-1} & p_{-2} \\ q_{-1} & q_{-2} \end{bmatrix} = \begin{bmatrix} 1 & 0 \\ 0 & 1 \end{bmatrix}$$

and (5.2.10) reduces to

$$\begin{bmatrix} p_k & p_{k-1} \\ q_k & q_{k-1} \end{bmatrix} = \begin{bmatrix} a_0 & 1 \\ 1 & 0 \end{bmatrix} \times \begin{bmatrix} a_1 & 1 \\ 1 & 0 \end{bmatrix} \times \cdots \times \begin{bmatrix} a_k & 1 \\ 1 & 0 \end{bmatrix}. \qquad (5.2.11)$$

Using this in (5.2.9) we finally derive that

$$\begin{bmatrix} p_k & p_{k-1} \\ q_k & q_{k-1} \end{bmatrix} \times \begin{bmatrix} r_{k-1} \\ r_k \end{bmatrix} = \begin{bmatrix} A \\ B \end{bmatrix}. \qquad (5.2.12)$$

Note that since the determinant of the product of any number of matrices is equal to the product of the determinants, from (5.2.11) we deduce that

$$p_k q_{k-1} - p_{k-1} q_k = (-1)^{k+1}$$

and this proves (5.2.3) (a). Now, if $r_n = 0$, then (5.2.12) for $k = n$ becomes

$$\begin{bmatrix} p_n & p_{n-1} \\ q_n & q_{n-1} \end{bmatrix} \times \begin{bmatrix} r_{n-1} \\ 0 \end{bmatrix} = \begin{bmatrix} A \\ B \end{bmatrix}$$

which simultaneously gives (5.2.3) (b) and (c) and our argument is complete.

Remark 5.1 Note that in the case that A and B are relatively prime the equation in (5.2.4) reduces to

$$1 = A (-1)^{n-1} q_{n-1} - B (-1)^n p_{n-1} . \qquad (\star)$$

This crucial identity, which here and after will be referred to as the *star identity*, makes the Euclidean algorithm a basic tool in the construction of inverses. To get across what we have in mind here we shall go over standard example. It is easy to verify that for any integer p the set of integers

$$\Omega_p = \{0, 1, 2, \ldots, p - 1\}$$

form a ring when multiplication and addition are carried out *mod p*. However, when p is a prime then every element in Ω_p, except 0, has an inverse. In fact, every one of these elements is prime with p, and the Euclidean algorithm carried out with A the given element and $B = p$ yields the identity

$$1 = A\,(-1)^{n-1} q_{n-1} \quad (\text{mod } p)$$

and we may set

$$A^{-1} = (-1)^{n-1} q_{n-1}.$$

This given it is easy to verify that the set Ω_p together with this notion of inverse is actually a field. Here and after we shall denote this field $\mathcal{F}_p$ and refer to it as a *prime field*. We should note that from the observations at the beginning of this section it follows that every field of characteristic p contains $\mathcal{F}_p$ as a subfield.

5.3   Polynomial Factorization

It is good to begin with the following basic fact.

Lemma 5.1 (Gauss) *If a polynomial* $P(x) = \sum_{m=0}^{n} p_m x^m$ *with integer coefficients factors over the rationals then it factors over the integers.*

Proof Clearly there is no loss in assuming that the coefficients of $P(x)$ have no common factor. Now if $P(x)$ factors over the rationals, we shall have two polynomials $A(x) = \sum_{i=0}^{\alpha_i} x^i$ and $B(x) = \sum_{j=0}^{b} b_i x^i$ with rational coefficients yielding $P(x) = A(x)B(x)$. Now let d_a be the least common multiple of the denominators of the a_i and d_b be the least common multiple of the denominators of the b_j. Let

$$R(x) = \sum_{i=0}^{a} r_i\, x^i = d_a A(x) \quad \text{and} \quad S(x) = \sum_{j=0}^{b} s_i\, x^i = d_b B(x).$$

Setting $M = d_a d_b$ we shall have the factorization

$$M P(x) = R(x)S(x) \tag{5.3.1}$$

where $R(x)$ and $S(x)$ are polynomials with integer coefficients. Note further that if the coefficients of $R(x)$ have a common prime factor then this factor must divide M. Likewise M must be divisible by any common prime factor of the coefficients of $S(x)$. By successive divisions of both sides of (5.3.1) by all such prime factors we can reduce (5.3.1) to an equation where both $R(x)$ and $S(x)$ have coefficients that have no common factor.

This given, let q be a prime factor of M. Equating constant terms in (5.3.1) gives

$$Mp_0 = r_0 s_0$$

and thus q must divide r_0 or s_0 or both. Equating coefficients of x in (5.3.1) gives

$$Mp_1 = r_0 s_1 + r_1 s_0 .$$

So if q did divide r_0 then q must divide $r_1 s_0$ and thus it must divide r_1, s_0 or both. From our assumptions q cannot divide all of the r_i's nor all of the s_j's. Let now i_0 be the first i such that q does not divide s_i and j_0 be the first j such that q does not divide s_j. But now let us equate coefficients of $x^{i_0+j_0}$ in (5.3.1). This gives

$$Mp_{i_0+j_0} = r_0 s_{i_0+j_0} + r_1 s_{i_0+j_0-1} + r_2 s_{i_0+j_0-2} + r_{i_0} s_{j_0} + \cdots$$

$$\cdots + r_{i_0+j_0-2} s_2 + r_{i_0+j_0-1} s_1 + r_{i_0+j_0} s_0$$

But we immediately see a contradiction here since the left hand side is divisible by q, all terms preceding $r_{i_0} s_{j_0}$ and all terms following $r_{i_0} s_{j_0}$ are divisible by q and by assumption $r_{i_0} s_{j_0}$ itself is not divisible by q. The contradiction arises from assuming that M is other than ± 1. This completes our proof. $\qquad\square$

Lemma 5.2 *In any field $\mathcal{F}$ a polynomial $P(x) \in \mathcal{F}[x]$ of degree n cannot have more than n roots.*

Proof Let

$$P(x) = c_0 + c_1 x + c_2 x^2 + \cdots + c_n x^n$$

then for any $\alpha \in \mathcal{F}$ we have

$$P(x) - P(\alpha) = \sum_{r=1}^{n} c_r \left(x^r - \alpha^r \right).$$

Since

$$x^r - \alpha^r = (x - \alpha) \sum_{s=0}^{r-1} x^s \alpha^{r-1-s}$$

it follows that we have the factorization

$$P(x) - P(\alpha) = (x - \alpha) Q(x) ,$$

Thus if $P(\alpha) = 0$ we have

$$P(x) = (x - \alpha)Q(x)$$

with $Q(x)$ a polynomial in $\mathcal{F}[x]$ of degree $n - 1$. This enables us to prove the assertion by induction on n. Indeed, for $n = 1$ this relation clearly shows that $P(x)$ can have at most 1 root in $\mathcal{F}$. But then if we inductively assume the Lemma to be true up to $n - 1$ then the same relation yields that $Q(x)$ cannot have more than $n - 1$ roots and thus $P(x)$ itself cannot have more than n roots completing the induction and the proof.   □

Lemma 5.3 *In any field $\mathcal{F}$ of order $> n$, a polynomial $P(x) = c_0 + c_1 x + \cdots + c_n x^n \in \mathcal{F}[x]$ satisfies*

$$P(\beta) = 0 \qquad \text{(for all } \beta \in \mathcal{F}\text{)}$$

if and only if

$$c_0 = c_1 = \cdots = c_n = 0.$$

Proof Since the *if* part is trivial we need only prove that the vanishing of P implies the vanishing of its coefficients. To this end note that by hypothesis $\mathcal{F}$ has at least $n + 1$ distinct elements

$$\beta_1, \beta_2, \ldots, \beta_{n+1}.$$

Evaluating P at these elements we get a system of equations

$$c_0 + c_1 \beta_1 + \cdots + c_n \beta_1^n = 0$$
$$c_0 + c_1 \beta_2 + \cdots + c_n \beta_2^n = 0$$
$$\vdots$$
$$c_0 + c_1 \beta_{n+1} + \cdots + c_n \beta_{n+1}^n = 0$$

with determinant

$$\prod_{1 \le i < j \le n+1} (\beta_i - \beta_j) \ne 0.$$

This forces the vanishing of the coefficients $c_0, c_1, \ldots, c_n$ and completes our proof.
                                                                                     □

Recall that a polynomial $P(x) \in \mathcal{F}[x]$ of positive degree is said to be *irreducible* in $\mathcal{F}$ if we cannot find two polynomials $R(x), S(x) \in \mathcal{F}[x]$ of degrees > 0 yielding the factorization

$$P(x) = R(x)S(x),$$

and called *monic* if its leading coefficient is 1.

Proposition 5.1 *Let* $A_1(x), A_2(x), \dots, A_k(x), \in \mathcal{F}[x]$ *and let* $P(x) \in \mathcal{F}[x]$ *irreducible in* $\mathcal{F}$. *If* $P(x)$ *divides the product*

$$A_1(x)A_2(x) \cdots A_k(x)$$

then it must necessarily divide at least one of the factors.

Proof Clearly there is nothing to prove if $k = 1$. So, proceeding by induction on k, let us assume the assertion valid up to $k - 1$. To prove it for k, for convenience set

$$A(x) = A_1(x), \qquad B(x) = A_2(x) \cdots A_k(x).$$

We need only to show that if $P(x)$ divides $A(x)B(x)$ and does not divide $A(x)$ then it must divide $B(x)$. This done the inductive hypothesis will yield the result for k. This given, suppose if possible that $P(x)$ does not divide $A(x)$ nor $B(x)$. Now note that the irreducibility of $P(x)$ forces the greatest common divisor of P and A to be either P itself or a scalar. The same holds true for P and B. Thus if P does not divide A nor B it follows from the equation in (5.2.4) of the Berlekamp algorithm that we can construct four polynomials $h_a, k_a, h_b, k_b \in \mathcal{F}[x]$ yielding the identities

$$1 = h_a(x)A(x) + k_a(x)P(x), \qquad 1 = h_b(x)B(x) + k_b(x)P(x).$$

But then multiplication of both sides gives

$$1 = h_a(x)h_b(x)A(x)B(x) + h_a(x)k_b(x)A(x)P(x)$$
$$+ k_a(x)h_b(x)P(x)B(x) + k_a(x)k_b(x)P(x)P(x)$$

and the assumption that $P(x)$ divides the product $A(x)B(x)$ immediately leads to a contradiction. This contradiction proves that if $P(x)$ does not divide $A(x)$ then it must divide $B(x)$ and our argument is complete.   □

This result has the following important implication.

Theorem 5.1 *Every monic polynomial* $P(x) \in \mathcal{F}[x]$ *has a unique factorization of the form*

$$P = \phi_1^{p_1}\phi_2^{p_2} \cdots \phi_k^{p_k} \tag{5.3.2}$$

with $\phi_1, \phi_2, \dots, \phi_k \in \mathcal{F}[x]$ *distinct, irreducible monic polynomials in* $\mathcal{F}[x]$.

Proof Note that if P is irreducible there is nothing to prove. On the other hand if $P(x)$ is not irreducible then it must have a factorization of the form $P(x) = A(x)B(x)$ with degree$(A) > 0$ and degree$(B) > 0$ and $A(x), B(x)$ monic polynomials in $\mathcal{F}[x]$. But then the identity

$$\text{degree}(A) + \text{degree}(B) = \text{degree}(P)$$

forces the inequalities

$$\text{degree}(A) < \text{degree}(P), \quad \text{degree}(B) < \text{degree}(P).$$

Applying the same reasoning on $A(x)$ and $B(x)$ and then to their factors, and then to the factors of their factors..., we can easily see that, (since the degrees decrease at each step), we will necessarily reach a factorization of the form

$$P(x) = A_1(x)A_2(x)\cdots A_n(x)$$

where each of the factors $A_i(x)$ is monic and irreducible in $\mathcal{F}[x]$. This proves existence since (5.3.2) can then be obtained by grouping together identical factors.

To show uniqueness, suppose if possible that we have

$$P = A_1(x)A_2(x)\cdots A_n(x) \text{ and } P = B_1(x)B_2(x)\cdots B_m(x) \qquad (5.3.3)$$

with all the $A_i(x)$ and $B_j(x)$ monic, irreducible polynomials in $\mathcal{F}[x]$.

We can immediately derive from these equalities that A_1 must be equal to one of the B_j. Indeed, from Proposition 5.1 and the irreducibility of A_1 and B_1, it follows that A_1 must be equal to B_1 or divide the product

$$B_2(x)B_3(x)\cdots B_m(x).$$

If $A_1 \neq B_1$ then the same reasoning will give that $A_1 = B_2$ or divide the product

$$B_3(x)\cdots B_m(x)$$

Clearly, proceeding in this manner we will necessarily find that $A_i = B_j$ for some $1 \leq j \leq m$. Thus, by relabeling the B_j, there is no loss in assuming that $A_1 = B_1$. This then yields the identity

$$A_2(x)\cdots A_n(x) = B_2(x)\cdots B_m(x).$$

It is easy to see that repeating this process we will eventually show that we must have $n = m$ and that in fact the two original factorizations in (5.3.3) where none other than one a permutation of the other. This proves uniqueness and completes our proof.   □

Remark 5.2 Note that if $f(x) \in \mathcal{F}_p[x]$ is a monic irreducible polynomial of degree k in $\mathcal{F}_p[x]$, then the collection of polynomials

$$\Omega_{f(x)} = \{c_0 + c_1x + \cdots + c_{k-1}x^{k-1} \mid c_i \in \mathcal{F}_p\}$$

form a field of order p^k. Here addition is ordinary addition of polynomials in $\mathcal{F}_p[x]$, multiplication is carried out *mod* $f(x)$ by the Division Algorithm and inverses are computed by means of the star identity of the Euclidean algorithm. We shall here and after denote this field by

$$\mathcal{F}_p[x]/(f(x))$$

and will refer to it as the *quotient field of* $f(x)$. Our ultimate goal is to show that every finite field may obtained by this construction and that for any integer $k \geq 1$, all quotient fields obtained from irreducible polynomials of degree k in this way are isomorphic.

5.4 Cyclotomic Polynomials

The classical Möbius function is defined by setting for any integer m

$$\mu(m) = \begin{cases} (-1)^k & \text{if } m \text{ is a product of } k \text{ distinct primes,} \\ 0 & \text{otherwise.} \end{cases} \tag{5.4.1}$$

For instance the following table gives first ten values of μ:

$$\begin{array}{rccccccccccc} m: & 1 & 2 & 3 & 4 & 5 & 6 & 7 & 8 & 9 & 10 \\ \mu(m): & 1 & -1 & -1 & 0 & -1 & 1 & -1 & 0 & 0 & 1 \end{array}$$

It is customary to define a partial order on the natural numbers by setting $d \preceq n$ if and only if d divides n. Sometimes the symbol "|" is used instead of $\preceq$.

The following basic identity is at the root of many uses of the Möbius function.

Proposition 5.2 *For any two integers $d \preceq n$ we have*

$$\sum_{d \preceq m \preceq n} \mu\left(\frac{n}{m}\right) = \begin{cases} 1 & \text{if } d = n, \\ 0 & \text{otherwise.} \end{cases} \tag{5.4.2}$$

Proof Note that if $d \preceq m \preceq n$ then we can write $m = dm'$ and $n = dn'$. This allows us to cancel the common factor d and derive that (5.4.2) is equivalent to

$$\sum_{1 \preceq m' \preceq n'} \mu\left(\frac{n'}{m'}\right) = \begin{cases} 1 & \text{if } n' = 1, \\ 0 & \text{otherwise.} \end{cases} \tag{5.4.3}$$

Clearly, when $d = n$ then $n' = 1$ and this sum reduces to the single term $\mu(1) = 1$. This gives the first case of (5.4.2). On the other hand when d is strictly less than n, then in (5.4.3) n'/m' runs over all divisors of n'. In other words we may as well rewrite (5.4.3) in the form

$$\sum_{1 \preceq m' \preceq n'} \mu(m') = \begin{cases} 1 & \text{if } n' = 1, \\ 0 & \text{otherwise.} \end{cases} \tag{5.4.4}$$

Now from the definition in (5.4.1) it follows that if $n' = p_1^{\alpha_1} p_2^{\alpha_2} \cdots p_k^{\alpha_k}$ then the only terms contributing to the sum in (5.4.4) are those where for some subset $S \subseteq \{1, 2, \ldots, k\}$ we have

$$m' = \prod_{i \in S} p_i$$

in which case $\mu(m') = (-1)^{|S|}$. Thus we are reduced to showing that

$$\sum_{S \subseteq \{1, 2, \ldots, k\}} (-1)^{|S|} = 0. \tag{5.4.5}$$

But this is immediate since in (5.4.5) the terms equal to 1 and those equal to -1 are equal in number. For instance, a bijection yielding this equality is given by the involution ϕ on subsets of $\{1, 2, \ldots, k\}$ obtained by letting $\phi(S) = S - \{1\}$ if $1 \in S$ and $\phi(S) = S + \{1\}$ if $1 \notin S$. This completes our argument. $\qquad\square$

This brings us the following important consequence of Proposition 5.2.

Theorem 5.2 *Two sequences of numbers $\{A_n\}_{n \geq 1}$ and $\{B_n\}_{n \geq 1}$ are related by the equations*

$$B_n = \sum_{d \preceq n} A_d \qquad (\text{for all } n \geq 1), \tag{5.4.6}$$

if and only if

$$A_n = \sum_{m \preceq n} B_m \, \mu\left(\frac{n}{m}\right) \qquad (\text{for all } n \geq 1). \tag{5.4.7}$$

Proof Note that if we set $B_m = \sum_{d \leq m} A_d$ in the right hand side of (5.4.7) it becomes

$$\sum_{m \leq n} \left(\sum_{d \leq m} A_d \right) \mu \left(\frac{n}{m} \right) .$$

Changing order of summation this can be rewritten as

$$\sum_d A_d \left(\sum_{d \leq m \leq n} \mu \left(\frac{n}{m} \right) \right)$$

and (5.4.7) immediately follows from (5.4.2).

Conversely, using (5.4.7), the right hand side of (5.4.6) becomes

$$\sum_{d \leq n} \sum_{m \leq d} B_m \, \mu \left(\frac{d}{m} \right) = \sum_m B_m \sum_d \mu \left(\frac{d}{m} \right) \chi \, (m \leq d \leq n) \qquad (5.4.8)$$

Now for a fixed m and n we have $m \leq d \leq n$ if and only if $d = d'm$ and $n = n'm$ thus

$$\sum_d \mu \left(\frac{d}{m} \right) \chi \, (m \leq d \leq n) = \sum_{d'} \mu(d') \chi \, (1 \leq d' \leq n')$$

and (5.4.4) gives

$$\sum_d \mu \left(\frac{d}{m} \right) \chi \, (m \leq d \leq n) = \begin{cases} 1 & \text{if } m = n, \\ 0 & \text{otherwise.} \end{cases}$$

Using this in (5.4.8) gives (5.4.6) and completes our proof.   □

Another important ingredient in the study of finite fields is the so called *Euler* Φ-*function*. This is the function $\Phi(n)$ which gives the number of integers in the interval $[1, n]$ that have no factor in common with n. In symbols, we may define $\Phi(n)$ by setting

$$\Phi(n) = \#\{ m \in [1, n] \mid \gcd(m, n) = 1 \} . \qquad (5.4.9)$$

It develops that $\Phi(n)$ has a most useful explicit formula.

Theorem 5.3 *For an integer with prime factorization*

$$n = p_1^{\alpha_1} p_2^{\alpha_2} \cdots p_k^{\alpha_k}$$

we have

$$\Phi(n) = n \left(1 - \frac{1}{p_1}\right) \left(1 - \frac{1}{p_2}\right) \cdots \left(1 - \frac{1}{p_k}\right) \tag{5.4.10}$$

as well as

$$\Phi(n) = \sum_{d \,\leq\, n} \mu(d) \frac{n}{d}. \tag{5.4.11}$$

Proof The best way to prove (5.4.10) is to use the *principle of inclusion exclusion* since by this path we gain one more tool for future use. The general set up is as follows. We have a finite set Ω and we are given certain subsets $A_1, A_2, \ldots, A_k \subseteq \Omega$. We are asked to count the number N of elements of Ω that do not belong to any of these subsets. Setting

$$\chi_{A_i}(x) = \begin{cases} 1 & \text{if } x \in A_i, \\ 0 & \text{otherwise.} \end{cases}$$

This number N is simply given by the expression

$$N = \sum_{x \in \Omega} \prod_{i=1}^{k} \left(1 - \chi_{A_i}(x)\right). \tag{5.4.12}$$

Since

$$\prod_{i=1}^{k} \left(1 - \chi_{A_i}(x)\right) = \sum_{S \subseteq [1,k]} (-1)^{|S|} \prod_{i \in S} \chi_{A_i}(x),$$

using this in (5.4.12) gives

$$N = \sum_{S \subseteq [1,k]} (-1)^{|S|} \sum_{x \in \Omega} \prod_{i \in S} \chi_{A_i}(x) \tag{5.4.13}$$

$$= \sum_{S \subseteq [1,k]} (-1)^{|S|} \# \bigcap_{i \in S} A_i.$$

Getting back to the proof of our theorem, if we let Ω be the set of integers in the interval $[1, n]$ and A_i be the subset of these integers that are divisible by p_i, then we can easily see that, with these choices, our $\Phi(n)$ is none other than the N in (5.4.12). To apply (5.4.13) we simply note that if $S = \{i_1, i_2, \ldots, i_s\}$ then, $\bigcap_{i \in S} A_i$ reduces to the subset of integers in $[1, n]$ that are divisible by the product $p_{i_1} p_{i_2} \cdots p_{i_s}$. Thus in this case, we have

$$\# \bigcap_{i \in S} A_i = \frac{n}{p_{i_1} p_{i_2} \cdots p_{i_s}}.$$

Using this in (5.4.13) we finally derive that

$$\Phi(n) = \sum_{s=0}^{n} \sum_{1 \le i_1 < i_2 < \cdots < i_s \le n} (-1)^s \frac{n}{p_{i_1} p_{i_2} \cdots p_{i_s}} \qquad (5.4.14)$$

and it is easily seen that this simply another way of writing (5.4.10). Moreover, since $\mu(p_{i_1} p_{i_2} \cdots p_{i_s}) = (-1)^s$ we see that (5.4.14) is also another way of writing (5.4.11). This completes our proof. $\qquad \square$

Recall that an element α of a group G is said to be *an nth root of unity* if $\alpha^n = 1$. We shall also say that α is of *order k* and write $ord(\alpha) = k$, if and only if

$$\alpha^k = 1 \quad \text{and} \quad \alpha^s \ne 1 \quad \text{for } 1 \le s \le k - 1.$$

There are four important properties of the notion of order which play an important role in our developments and are good to keep in mind.

Proposition 5.3 *In a multiplicative group G:*

(1) An element $\alpha \in G$ has order n then $\alpha^m = 1$ if and only if n divides m.
(2) If an element $\alpha \in G$ has order n then the element α^k has order $\frac{n}{\gcd(n,k)}$
(3) If $\alpha, \beta \in G$ have orders a and b respectively and $\gcd(a, b) = 1$ then $\alpha\beta$ has order ab.
(4) If n is the maximal order of any element of G then the order of any other element of G is a divisor of n.

Proof Note that if $ord(\alpha) = n$ and $\alpha^m = 1$ then of course we must have $m \ge n$ and if $m > n$ then by division we get $m = nQ + R$ with $0 \le R < n$ but then

$$\alpha^m = \alpha^{nQ+R} = \alpha^R = 1$$

and $R > 0$ contradicts the nature of n. This proves (1). Now again if $ord(\alpha) = n$ then we see that

$$(\alpha^k)^{\frac{n}{\gcd(n,k)}} = (\alpha^n)^{\frac{k}{\gcd(n,k)}} = 1.$$

On the other hand if $(\alpha^k)^m = 1$ then from (1) it follows that n divides km so suppose that for some integer r we have $km = rn$. But we must also have

$$\frac{k}{\gcd(n,k)} m = r \frac{n}{\gcd(n,k)}$$

and since $\frac{k}{\gcd(n,k)}$ and $\frac{n}{\gcd(n,k)}$ are relatively prime it follows from this that $\frac{n}{\gcd(n,k)}$ must divide m. This proves (2). To prove (3) note that

$$(\alpha\beta)^m = 1 \quad \rightarrow \quad \alpha^m = \beta^{-m} \quad \text{and} \quad \beta^m = \alpha^{-m}$$

and thus we also have

$$\alpha^{mb} = 1 \quad \text{and} \quad \beta^{ma} = 1$$

and (1) together with $ord(\alpha) = a$ and $ord(\beta) = b$ give that

$$a | mb \quad \text{and} \quad b | ma .$$

But now the hypothesis that $\gcd(a, b) = 1$ forces m to be divisible by the product ab. Since we clearly have

$$(\alpha\beta)^{ab} = (\alpha^a)^b (\beta^b)^a = 1$$

our proof of (3) is now complete.

Finally, suppose that $\alpha \in G$ has order n and every other element of G has order $\le n$. Suppose if possible that some element $\beta \in G$ has order m and that m does not divide n. Then there will be some prime factor p of m with the property that p^e (for some $e \ge 1$) divides m and p^{e-1} is the maximal power of p that divides n. Now from (2) it follows that

$$ord\left(\alpha^{p^{e-1}}\right) = \frac{n}{p^{e-1}} \quad \text{and} \quad ord\left(\beta^{\frac{m}{p^e}}\right) = p^e .$$

Since $\frac{n}{p^{e-1}}$ and p^e are relatively prime it follows from (3) that

$$ord\left(\alpha^{p^{e-1}} \beta^{\frac{m}{p^e}}\right) = n\,p$$

and that contradicts that n is the largest order of any element of G. This completes our proof.   □

Now for some integer $n > 1$, set $\omega = e^{2\pi i/n}$ and note that the n complex numbers

$$1, \omega, \omega^2, \ldots, \omega^{n-1} \tag{5.4.15}$$

are all solutions of the equation $t^n - 1 = 0$ and since they are distinct and the polynomial $1 - t^n$ has at most n roots it follows that we must have

$$t^n - 1 = \prod_{i=0}^{n-1} (t - \omega^i). \tag{5.4.16}$$

Note further that if d is a divisor of n then by (1) of Proposition 5.3 it follows that every complex number β of order d must be a root of the polynomial $t^n - 1$ and therefore must be one of the elements in (5.4.15). Thus if we set

$$Q_d(t) = \prod_{\substack{ord(\beta)=d \\ \beta \in \beta}} (t - \beta) \tag{5.4.17}$$

then we must necessarily have the factorization

$$t^n - 1 = \prod_{d|n} Q_d(t) . \tag{5.4.18}$$

The expression in (5.4.17) defines the so called dth *cyclotomic polynomial*. As we shall see this polynomial has remarkable properties and plays a basic role in every field. To begin with it can be given an explicit formula.

Theorem 5.4 *For all $d > 1$ we have*

$$Q_m(t) = \prod_{n|m} (t^n - 1)^{\mu(\frac{m}{n})} . \tag{5.4.19}$$

In particular $Q_m(t)$ is always a monic polynomial with integer coefficients.

Proof Note that using (5.4.18) in the right hand side of (5.4.19) we get

$$RHS = \prod_{n|m} \prod_{d|n} Q_d(t)^{\mu(\frac{m}{n})}$$

$$= \prod_d Q_d(t)^{\sum_{d \le n \le m} \mu(\frac{m}{n})}$$

and (5.4.19) follows by another use of (5.4.2). So we can write (5.4.19) in the form

$$Q_m(t) \prod_{\substack{n|m \\ \mu(m/n)=-1}} (t^n - 1) = \prod_{\substack{n|m \\ \mu(m/n)=1}} (t^n - 1) . \tag{5.4.20}$$

Now both products on the right and on the left of this equality are clearly monic polynomials with integer coefficients. We thus deduce from (5.4.20), by the division algorithm, that $Q_m(t)$ must be a monic polynomial with rational coefficients and thus the last assertion is an immediate consequence of Gauss lemma. This completes our proof.   □

Formula (5.4.19) gives a practical way to compute a given cyclotomic polynomial. For instance, when $m = 6$ from the following table

n	1	2	3	6
$\frac{m}{n}$	6	3	2	1
$\mu(\frac{m}{n})$	1	-1	-1	1
$x^n - 1$	$x - 1$	$x^2 - 1$	$x^3 - 1$	$x^6 - 1$

we get

$$Q_6(x) = \frac{(x - 1)(1 - x^6)}{(x^2 - 1)(x^3 - 1)} = \frac{(x^3 + 1)}{(x + 1)} = 1 - x + x^2 .$$

We give below a list of the polynomials $Q_{p-1}(x)$ as p runs over the first 14 primes.

$$Q_1(x) = x - 1$$
$$Q_2(x) = 1 + x$$
$$Q_4(x) = 1 + x^2$$
$$Q_6(x) = 1 - x + x^2$$
$$Q_{10}(x) = 1 - x + x^2 - x^3 + x^4$$
$$Q_{12}(x) = 1 - x^2 + x^4$$
$$Q_{16}(x) = 1 + x^8$$
$$Q_{18}(x) = 1 - x^3 + x^6$$
$$Q_{22}(x) = 1 - x + x^2 - x^3 + x^4 - x^5 + x^6 - x^7 + x^8 - x^9 + x^{10}$$
$$Q_{28}(x) = 1 - x^2 + x^4 - x^6 + x^8 - x^{10} + x^{12}$$
$$Q_{30}(x) = 1 + x - x^3 - x^4 - x^5 + x^7 + x^8$$
$$Q_{36}(x) = 1 - x^6 + x^{12}$$
$$Q_{40}(x) = 1 - x^4 + x^8 - x^{12} + x^{16}$$
$$Q_{42}(x) = 1 + x - x^3 - x^4 + x^6 - x^8 - x^9 + x^{11} + x^{12}$$

We should point out the following important relation between the Euler Φ-function and the cyclotomic polynomials.

Proposition 5.4

$$\text{degree } Q_m(x) = \Phi(m) , \tag{5.4.21}$$

and in particular by (5.4.18)

$$\sum_{d \preceq n} \Phi(d) = n . \tag{5.4.22}$$

Proof From (5.4.20) it follows that the degree of the polynomial $Q_n(x)$ may be computed by subtracting the degree of the product on the left from the degree of the product on the right. This gives

$$\text{degree } Q_m(x) = \sum_{\substack{n|m \\ \mu(m/n)=1}} n - \prod_{\substack{n|m \\ \mu(m/n)=-1}} n = \sum_{n \preceq m} n \, \mu\left(\frac{m}{n}\right) .$$

Thus (5.4.21) follows from the identity in (5.4.11). We also see from formula (5.4.18) that n, which is the degree of $1 - x^n$, must also be equal to the sum of the degrees of the factors $\phi_d(x)$ as d varies over the divisors of n. This establishes (5.4.22) and completes our proof. □

In a finite field $\mathcal{F}$ of order q an element α is said to be *primitive* if and only if

$$(i) \quad \alpha^{q-1} = 1 \quad \text{and} \tag{5.4.23}$$

$$(ii) \quad \{\alpha, \alpha^2, \dots, \alpha^{q-1}\} = \mathcal{F} - \{0\} .$$

Note all this says, in this case, is that $ord(\alpha) = q - 1$.

Propositions 5.3 and 5.4 combined have the following surprising corollary.

Theorem 5.5 *A finite field $\mathcal{F}$ of order q has exactly $\Phi(q - 1)$ primitive elements. In fact, for any $d \preceq q - 1$, the cyclotomic polynomial $Q_d(t)$ has exactly $\Phi(d)$ roots in $\mathcal{F}$ and they constitute all the elements of order d in $\mathcal{F}$.*

Proof Since $\mathcal{F}$ has order $q < \infty$ every nonzero element $\beta \in \mathcal{F}$ has an order bounded by $q - 1$. Note that if we set

$$n_0 = \max\{ord(\beta) \mid \beta \in \mathcal{F}\}$$

then from (4) of Proposition 5.3 it follows that in the multiplicative group $\mathcal{F} - \{0\}$ the order of every element must divide n_0. But that means that the elements of $\mathcal{F} - \{0\}$ are all roots of the equation

$$x^{n_0} - 1 = 0 .$$

Since from Lemma 5.2 this equation cannot have more than n_0 roots it follows that

$$\#(\mathcal{F} - \{0\}) \leq n_0 .$$

This gives

$$q - 1 \leq n_0 .\tag{5.4.24}$$

On the other hand if α has order n_0 then the elements

$$1, \alpha, \alpha^2, \ldots, \alpha^{n_0-1}\tag{5.4.25}$$

are distinct and since they are all in $\mathcal{F} - \{0\}$ we must necessarily also have

$$n_0 \leq q - 1.$$

This together with (5.4.24) proves the equality

$$n_0 = q - 1$$

and from the definition in (5.4.23) it follows that α is a primitive element of $\mathcal{F}$. Moreover this equality now yields the following three basic facts:

(1) Every nonzero element of $\mathcal{F}$ must have an order that is a divisor of $q - 1$,
(2) Every nonzero element of $\mathcal{F}$ must be one of the elements in (5.4.25),
(3) Every nonzero element of $\mathcal{F}$ must be a root of the polynomial $x^{q-1} - 1$.

Note that since $x^{q-1} - 1$ has no more than $q - 1$ roots it follows from (2) and (3) that

$$x^{q-1} - 1 = \prod_{\beta \in \mathcal{F}-\{0\}} (x - \beta) = \prod_{i=0}^{q-2}(x - \alpha^i) .\tag{5.4.26}$$

But now (1) combined with the first equality in (5.4.26) yields that we must have

$$x^{q-1} - 1 = \prod_{\substack{d \preceq q-1}} \prod_{\substack{\beta \in \mathcal{F}-\{0\} \\ ord(\beta)=d}} (x - \beta).\tag{5.4.27}$$

Now fix $n \preceq q - 1$ and set $\gamma = \alpha^{\frac{q-1}{n}}$. By (2) of Proposition 5.3 it follows that $ord(\gamma) = n$, and the thus the elements

$$1, \gamma, \gamma^2, \ldots, \gamma^{n-1}$$

are distinct and satisfy the equation $x^n - 1 = 0$. Since this equation cannot have more than n solutions in $\mathcal{F}$, it follows that these elements are all the nth roots of unity of $\mathcal{F}$. Thus we must have the two factorizations

$$x^n - 1 = \prod_{i=0}^{n-1}(x - \gamma^i) = \prod_{\substack{\beta\in\mathcal{F}-\{0\} \\ \beta^n=1}} (x - \beta).$$

Grouping the factors of the last product according to the order of the corresponding element we derive that for any $n \preceq q - 1$ we must also have

$$x^n - 1 = \prod_{d \preceq n} \prod_{\substack{\beta\in\mathcal{F}-\{0\} \\ ord(\beta)=d}} (x - \beta). \tag{5.4.28}$$

Setting for a moment

$$R_d(x) = \prod_{\substack{\beta\in\mathcal{F}-\{0\} \\ ord(\beta)=d}} (x - \beta)$$

(5.4.28) becomes

$$x^n - 1 = \prod_{d \preceq n} R_d(x) \qquad \text{(for all } n \preceq q - 1).$$

Thus for any $m \preceq q - 1$ we have

$$\prod_{n \preceq m} (x^n - 1)^{\mu(\frac{m}{n})} = \prod_{n \preceq m} \prod_{d \preceq n} \left(R_d(x)\right)^{\mu(\frac{m}{n})}$$

$$= \prod_{d} \left(R_d(x)\right)^{\sum_{d \preceq n \preceq m} \mu(\frac{m}{n})}$$

$$\text{(by (5.4.2))} = R_m(x)$$

and (5.4.19) gives

$$R_m(x) = Q_m(x).$$

Thus the factorization in (5.4.27) is none other than

$$x^{q-1} - 1 = \prod_{d \preceq q-1} Q_d(x). \tag{5.4.29}$$

Now note that the polynomial $1 - x^{q-1}$ has $q - 1$ roots given by the elements

$$1, \alpha, \alpha^2, \ldots, \alpha^{q-2}. \tag{5.4.30}$$

Moreover (5.4.22) for $n = q - 1$ gives

$$\sum_{d \preceq q-1} \Phi(d) = q - 1 .$$

Thus from (5.4.29), (5.4.21) and Lemma 5.2 it follows that for $d \preceq q - 1$ each polynomial $Q_d(x)$ has precisely $\Phi(d)$ roots in $\mathcal{F}$ and these roots are all the elements of order d in $\mathcal{F}$. This proves the last assertion of the Theorem. In particular we derive that $\mathcal{F}$ contains exactly $\Phi(q - 1)$ elements of order $q - 1$ which a fortiori must be all the primitive roots of $\mathcal{F}$. This completes our proof.   □

We should note that the proof of Theorem 5.5 yields also the following interesting fact.

Proposition 5.5 *If α is any primitive root of $\mathcal{F}$ then the set of elements*

$$\mathcal{P} = \{\alpha^k \mid \gcd(k, q - 1) = 1\}$$

gives all the primitive roots of $\mathcal{F}$. In particular we also have

$$Q_{q-1}(x) = \prod_{\substack{1 \le k \le q-1 \\ \gcd(k, q-1)=1}} (x - \alpha^k) . \qquad (5.4.31)$$

Proof It follows from (2) of Proposition 5.3 that all the elements of $\mathcal{P}$ have order $q - 1$. On the other hand the definition in (5.4.9) of the Euler Φ-function gives that the set $\mathcal{P}$ has cardinality precisely $\Phi(q - 1)$ but the latter is, in fact, also the degree of $Q_{q-1}(x)$ whose roots are all the primitive elements of $\mathcal{F}$. This proves the Proposition (including (5.4.31)).   □

Problems

1. Since it follows from the Euclidean algorithm that the integers modulo a prime p form a field $\mathcal{F}_p$. We can use Theorem 5.4 to determine the primitive roots of $\mathcal{F}_p$ for any prime p. Test your understanding of the material in this subsection by constructing all the primitive roots of $\mathcal{F}_{19}$.
2. Use the results of Problem 1. to construct the inverse of 15 in $\mathcal{F}_{19}$ without using the Euclidean algorithm.

5.5   The Frobenius Map

In a finite field $\mathcal{F}$ of characteristic p the map $\phi : \mathcal{F} \to \mathcal{F}$ obtained by setting

$$\phi(\beta) = \beta^p \qquad \text{(for all } \beta \in \mathcal{F}) \tag{5.5.1}$$

is usually referred to as the *Frobenius map*. As we shall soon see, this map has a variety of truly remarkable properties, but before we can present them we need some auxiliary facts. To begin we shall assume, here and after, whether explicitly stated or not, that our field $\mathcal{F}$ has characteristic p.

Proposition 5.6 *For any elements $a_1, a_2, \ldots, a_n \in \mathcal{F}$ and any integer $k > 0$ we have*

$$(a_1 + a_2 + \cdots + a_n)^{p^k} = a_1^{p^k} + a_2^{p^k} + \cdots + a_n^{p^k}. \tag{5.5.2}$$

Proof To begin note that the binomial coefficient $\binom{p}{k}$ necessarily vanishes in $\mathcal{F}$ for any $r = 1, 2, \ldots, p - 1$. Indeed we see that, p being a prime, for these values of r, the factors in the denominators in the formula

$$\binom{p}{r} = \frac{p!}{r!(p-r)!}$$

cannot cancel the factor p contained in $p!$. This given, the identity

$$(a_1 + a_2)^p = \sum_{r=0}^{p} \binom{p}{r} a_1^r a_2^{p-r}$$

reduces to

$$(a_1 + a_2)^p = a_1^p + a_2^p \qquad \text{(in } \mathcal{F}).$$

This proves (5.5.2) for $n = 2$ and $k = 1$. To prove the general case assume by induction that (5.5.2) is true for $k = 1$ and $n - 1$. But then the $n = 2$ case and the inductive hypothesis gives

$$(a_1 + a_2 + \cdots + a_n)^p = a_1^p + (a_2 + \cdots + a_n)^p = a_1^p + a_2^p + \cdots + a_n^p.$$

This completes the induction and proves (5.5.2) for $k = 1$. Next note that a double use of $k = 1$ case identity gives

$$(a_1 + a_2 + \cdots + a_n)^{p^2} = (a_1^p + a_2^p + \cdots + a_n^p)^p = a_1^{p^2} + a_2^{p^2} + \cdots + a_n^{p^2}$$

and we can easily see that what remains to be proved follows again by an inductive argument. □

This identity has an immediate important corollary.

Proposition 5.7 *In a field $\mathcal{F}$ of characteristic p no element can have an order that is divisible by p.*

Proof Suppose that for some element $\beta \in \mathcal{F}$ we have

$$\beta^{mp} = 1 .$$

Then since (5.5.2) gives

$$(\beta^m - 1)^p = \beta^{mp} + (-1)^p = \beta^{mp} - 1 .$$

This is also true when $p = 2$ since in characteristic 2 we have $1 = -1$. We must also have

$$\beta^m = 1 .$$

Thus mp cannot possibly be the order of β. This completes our argument. □

Proposition 5.8 *An element $\alpha \in \mathcal{F}$ belongs to the subfield $\mathcal{F}_p$ if and only if*

$$\alpha^p = \alpha . \tag{5.5.3}$$

In other words under the Frobenius map the only elements of $\mathcal{F}$ that remain fixed are those in $\mathcal{F}_p$.

Proof Since $\mathcal{F}_p$ has p elements it follows from the proof of Theorem 5.5 that all the nonzero elements of $\mathcal{F}_p$ satisfy the polynomial equation

$$x^{p-1} - 1 = 0$$

and this means that $0, 1, \ldots, p - 1$ satisfy the equation

$$x^p - x = 0 .$$

Since the polynomial $x^p - x$ cannot have more than p roots in $\mathcal{F}$, the elements of $\mathcal{F}_p$ must be the only ones in $\mathcal{F}$ that satisfy the condition in (5.5.3). This completes the proof. □

We are now in a position to establish a most remarkable property of finite fields.

Theorem 5.6 *In a finite field $\mathcal{F}$ of characteristic p an element $\beta \in \mathcal{F}$ of order d generates, under the action of the Frobenius map ϕ, a cycle of length m, where m is the order of p in the integers mod d. In other words the elements*

$$\beta, \beta^p, \beta^{p^2}, \ldots, \beta^{p^{m-1}} \tag{5.5.4}$$

are distinct and we have

$$\beta^{p^m} = \beta. \tag{5.5.5}$$

Thus, under ϕ, $\mathcal{F}$ breaks up into the union of disjoint cycles. In other words the Frobenius map permutes the elements of $\mathcal{F}$ and therefore it is invertible in $\mathcal{F}$.

Proof By definition, m is the order of p mod d if and only if

$$(a) \ p^m = 1 \ (\text{mod } (d) \ \text{and} \ \ (b) \ p^s \neq 1 \ (\text{mod } (d) \ (\text{for } 1 \leq s \leq m - 1). \tag{5.5.6}$$

Clearly, (5.5.6) (a) gives (5.5.5). Moreover, if for some pair $i < j$ we have

$$\beta^{p^j} = \beta^{p^i}$$

then dividing both sides by β^{p^i} we get

$$\beta^{p^i(p^{j-i}-1)} = 1.$$

In particular, d must divide the product $p^i(p^{j-i} - 1)$. Since by Proposition 5.7 d cannot have p as a factor, it follows that d must divide $p^{j-i} - 1$. In other words

$$p^{j-i} = 1 \quad (\text{mod } (d).$$

But from $b)$ of (5.5.6) we see that this cannot happen when $1 \leq i < j \leq m - 1$. Thus the elements in (5.5.4) are distinct, precisely as asserted, and (5.5.5) shows that these elements constitute a cycle of the Frobenius map. Since every element of $\mathcal{F}$ has a finite order and cycles that share an element are necessarily identical, the last two assertions are immediate. $\qquad\qquad\qquad\square$

Proposition 5.9 *In a finite field $\mathcal{F}$ of order q, polynomial $P(x) \in \mathcal{F}[x]$ of degree $n < q$ satisfies the identity*

$$P(\beta)^p = P(\beta^p) \qquad (\text{for all } \beta \in \mathcal{F}) \tag{5.5.7}$$

if and only if

$$P(x) \in \mathcal{F}_p[x].$$

Proof Let

$$P(x) = c_0 + c_1 x + \cdots + c_n x^n .$$

We are to show that (5.5.7) holds if and only if

$$c_0 \in \mathcal{F}_p , \quad c_1 \in \mathcal{F}_p , \quad \ldots , \quad c_n \in \mathcal{F}_p . \tag{5.5.8}$$

Note first that from (5.5.2) it follows that

$$P(\beta)^p = c_0^p + c_1^p \beta^p + \cdots + c_n^p \beta^{pn} .$$

Thus the condition in (5.5.7) is none other than

$$c_0^p + c_1^p \beta^p + \cdots + c_n^p \beta^{pn} = c_0 + c_1 \beta^p + \cdots + c_n \beta^{pn} \tag{5.5.9}$$

that is we must have

$$Q(x) = c_0^p - c_0 + (c_1^p - c_1)\beta^p + \cdots + (c_n^p - c_n)\beta^{pn} = 0 \quad \text{(for all } \beta \in \mathcal{F}) . \tag{5.5.10}$$

Since by Theorem 5.6 the Frobenius map permutes the elements of $\mathcal{F}$ it follows that (5.5.10) is equivalent to the condition

$$Q(x) = c_0^p - c_0 + (c_1^p - c_1)\beta + \cdots + (c_n^p - c_n)\beta^n = 0 \quad \text{(for all } \beta \in \mathcal{F}) .$$

However since $n < q$ we can use Lemma 5.3 and derive that

$$c_0^p = c_0 , \quad c_1^p = c_1 , \quad \ldots , \quad c_n^p = c_n \tag{5.5.11}$$

and (5.5.8) then follows from Proposition 5.8. Conversely, by the same Proposition, the relations in (5.5.8) give (5.5.11) and (5.5.10) must hold true for all $\beta \in \mathcal{F}$ and we have seen that this is equivalent to (5.5.7). Thus our proof is complete.   □

This brings us to another basic result on finite fields.

Theorem 5.7 *In a finite field $\mathcal{F}$ of order q and characteristic p an element β of order d is a root of a monic irreducible polynomial $\phi_\beta(x) \in \mathcal{F}_p[x]$ of degree equal to the order of p in the integers mod d. In fact, we have*

$$\phi_\beta(x) = \prod_{s=0}^{m-1} (x - \beta^{p^s}) \tag{5.5.12}$$

where $\beta, \beta^p, \cdots, \beta^{p^{m-1}}$ are none other than the elements of the Frobenius cycle generated by β. It then follows that β generates within $\mathcal{F}$ a subfield of order p^m that is isomorphic to the quotient field $\mathcal{F}_p[x]/(\phi_\beta(x))$.

Proof To prove the first assertion it suffices to show that the polynomial $\phi_\beta(x)$ defined by (5.5.12) lies in $\mathcal{F}_p[x]$ and is irreducible in $\mathcal{F}_p[x]$. To this end note first that since by Proposition 5.6 we have

$$(x - \beta^{p^s})^p = (x^p - \beta^{p^{s+1}})$$

we derive that

$$\phi_\beta(x)^p = \prod_{s=1}^m (x^p - \beta^{p^s}) = \phi_\beta(x^p) .$$

Since $m < q$ we can use Proposition 5.9 and derive that $\phi_\beta(x) \in \mathcal{F}_p[x]$. Next suppose that for two polynomials $A(x), B(x) \in \mathcal{F}_p[x]$ of degree $< m$ we have

$$A(x)B(x) = \phi_\beta(x)$$

then setting $x = \beta$ gives

$$A(\beta)B(\beta) = \phi_\beta(\beta) = 0.$$

Thus if $B(\beta) \neq 0$ we must have $A(\beta) = 0$, but then from Proposition 5.9 we derive that

$$A(\beta) = A(\beta^p) = A(\beta^{p^2}) = \cdots = A(\beta^{p^{m-1}}) = 0$$

but this yields a contradiction since a polynomial of degree $< m$ cannot have m distinct roots. The same contradiction arises if we assume that $A(\beta) \neq 0$. This proves that $\phi_\beta(x)$ is irreducible in $\mathcal{F}_p[x]$.

Now note that the collection

$$\mathcal{F}_p[\beta] = \left\{ c_0 + c_1\beta + \cdots + c_{m-1}\beta^{m-1} \mid c_0, c_1, \ldots, c_{m-1} \in \mathcal{F}_p \right\}$$

consists of p^m distinct elements. Indeed, the equality of any two of them yields a polynomial equation of degree less than m satisfied by β, contradicting the irreducibility of $\phi_\beta(x)$. Furthermore, it is easily seen that $\mathcal{F}_p[\beta]$ is a subfield of $\mathcal{F}$ which is isomorphic to the quotient field $\mathcal{F}[x]/(\phi_\beta(x))$. In fact the isomorphism is simply given by the map

$$ev_\beta : \mathcal{F}[x]/(\phi_\beta(x)) \to \mathcal{F}_p[\beta]$$

obtained by setting

$$ev_\beta f(x) = f(\beta) .$$

This map is clearly onto and well defined since if for three elements $f(x), g(x), h(x) \in \mathcal{F}_p[x]$ we have

$$f(x) = g(x) + \phi_\beta(x)h(x)$$

then then setting $x = \beta$ gives

$$f(\beta) = g(\beta).$$

It is also injective since if for some $f(x) \in \mathcal{F}_p[x]$ we have

$$f(\beta) = 0$$

then the irreducibility of $\phi_\beta(x)$ yields that $f(x)$ must be a multiple of $\phi_\beta(x)$. Likewise, ev_β preserves multiplication since if for four elements $f(x), g(x), h(x), k(x) \in \mathcal{F}_p[x]$ we have

$$f(x)g(x) = h(x) + \phi_\beta(x)k(x)$$

then setting $x = \beta$ gives

$$f(\beta)g(\beta) = h(\beta).$$

Similarly if

$$f(x) + g(x) = h(x) + \phi_\beta(x)k(x),$$

then setting $x = \beta$ gives

$$f(\beta) + g(\beta) = h(\beta).$$

Thus ev_β preserves addition as well and our proof is complete.   □

An immediate corollary of Theorem 5.7 yields the nature of a finite field.

Theorem 5.8 *Let be a finite field $\mathcal{F}$ of characteristic p and order q then*

$$q = p^k \tag{5.5.13}$$

where k is the order of p mod $q - 1$. In fact, for any primitive element $\alpha \in \mathcal{F}$ we have the equality

$$\mathcal{F}_p[\alpha] = \mathcal{F}. \tag{5.5.14}$$

Proof Theorem 5.5 gives that $\mathcal{F}$ has $\Phi(q-1)$ primitive elements. Let α be one of them. From Theorem 5.7 it follows that α generates a subfield of

$$\mathcal{F}_p[\alpha] \subseteq \mathcal{F} \qquad (5.5.15)$$

of order p^k with k the order of p mod $q-1$. Thus

$$p^k \le q. \qquad (5.5.16)$$

On the other hand since we must have $p^k - 1 = 0$ modulo $q-1$ it follows that $q-1$ must divide $p^k - 1$. This gives

$$q - 1 \le p^k - 1 \qquad (5.5.17)$$

and the equality in (5.5.13) immediately follows by combining (5.5.16) and (5.5.17). But now (5.5.13) together with (5.5.15 forces the equality in (5.5.14) and completes our proof.   □

Theorem 5.8 essentially states that every finite field of characteristic p is isomorphic to a quotient field $\mathcal{F}_p[x]/(f(x))$ with $f(x)$ a monic polynomial irreducible in $\in \mathcal{F}_p[x]$. To get deeper into the nature of finite fields we need two auxiliary results.

Lemma 5.4 *For any given prime p and any two integers $1 \le h < k$ we have*

$$p^h - 1 \preceq p^k - 1 \qquad \Longleftrightarrow \qquad h \preceq k \qquad (5.5.18)$$

Proof It easily verified that for any three integers h, k, r we have

$$x^k - 1 = (x^h - 1)\left(1 + x^h + x^{2h} + \cdots + x^{(r-1)h}\right) + x^{rh}\left(x^{k-rh} - 1\right) \qquad (5.5.19)$$

We shall use it here for $h < k$ and

$$r = \max\{i \mid k - ih \ge 0\}, \qquad (5.5.20)$$

and note that we have

$$h \preceq k \qquad \Longleftrightarrow \qquad k - rh = 0. \qquad (5.5.21)$$

Making the replacement $x \to p$ in (5.5.19) gives

$$p^k - 1 = (p^h - 1)\left(1 + p^h + p^{2h} + \cdots + p^{(r-1)h}\right) + p^{rh}\left(p^{k-rh} - 1\right). \qquad (5.5.22)$$

Note that if $h \preceq k$ then from (5.5.21) it follows that

$$p^k - 1 = (p^h - 1)\left(1 + p^h + p^{2h} + \cdots + p^{(r-1)h}\right),$$

which yields that $p^h - 1$ divides $p^k - 1$. This proves the left pointing arrow in (5.5.18). To prove the converse we need only show that $p^h - 1 \preceq p^k - 1$ forces $k - rh = 0$. To this end suppose if possible that $k - rh > 0$ and note that if $p^h - 1$ divides $p^k - 1$ then it follows from (5.5.22) that $p^h - 1$ must also divide $p^{rh}(p^{k-rh} - 1)$. Since $p^h - 1$ can't have any factor in common with p^{rh} it must be that $p^h - 1$ divides $p^{k-rh} - 1$. Now this implies

$$p^h - 1 \le p^{k-rh} - 1$$

or equivalently

$$h \le k - rh.$$

But this is in plain contradiction with the definition of r in (5.5.20). Thus only the alternative $k - rh = 0$ is compatible with $p^h - 1 \preceq p^k - 1$ and our proof is complete. $\square$

The equivalence in (5.5.18) has a most unexpected consequence.

Theorem 5.9 *In a finite field $\mathcal{F}$ of order p^k all the Frobenius cycles have lengths that divide k.*

Proof Let β be an element of $\mathcal{F}$ that generates a cycle of length h. Now it follows from Theorem 5.7 that β generates a subfield $\mathcal{F}_p(\beta) \subseteq \mathcal{F}$ of order p^h. Applying Theorem 5.5 to $\mathcal{F}_p(\beta)$ we derive that $\mathcal{F}_p(\beta)$ (and therefore $\mathcal{F}$) contains an element α of order $p^h - 1$. But from the proof of Theorem 5.5 it follows that all the elements of $\mathcal{F}$ have an order that divides $p^k - 1$. In particular we must have $p^h - 1 \preceq p^k - 1$ and Lemmma 5.4 yields that $h \preceq k$ completing our proof. $\square$

Proposition 5.10 *Let $\mathcal{F}$ be a field of order q and let $I_d(\mathcal{F})$ denote the number of monic polynomials of degree d in $\mathcal{F}[x]$ which are irreducible in $\mathcal{F}[x]$. Then*

$$q^n = \sum_{d \preceq n} d I_d(\mathcal{F}) \qquad (\text{for all } n \ge 1) \tag{5.5.23}$$

and therefore from Theorem 5.1 it follows that

$$I_m(\mathcal{F}) = \frac{1}{m} \sum_{n \preceq m} q^n \mu\left(\frac{m}{n}\right). \tag{5.5.24}$$

Proof Let $\mathcal{M}(\mathcal{F})$ denote the collection of all monic polynomials in $\mathcal{F}[x]$ and let $\mathcal{IM}(\mathcal{F})$ be the subcollection of irreducible ones. Note that if

$$\mathcal{IM}(\mathcal{F}) = \{\phi_1, \phi_2, \phi_3, \ldots, \phi_r, \ldots\}$$

then, by Theorem 5.1 every $P \in \mathcal{M}(\mathcal{F})$ may be uniquely expressed in the form

$$P = \phi_{i_1}^\alpha \phi_{i_2}^\alpha \cdots \phi_{i_k}^\alpha \qquad (\text{for some } 1 \le i_1 < i_2 < \cdots < i_k).$$

In other words every $P \in \mathcal{M}(\mathcal{F})$ may be viewed as a monomial in the infinite set of variables $\phi_1, \phi_2, \phi_3, \ldots, \phi_r, \ldots$. In this vein, the unique factorization result may simply be translated into the formal power series identity

$$\sum_{P \in \mathcal{M}(\mathcal{F})} P = \prod_{i \geq 1}(1 + \phi_i + \phi_i^2 + \cdots) = \prod_{i \geq 1} \frac{1}{1 - \phi_i}. \qquad (5.5.25)$$

Now replacing each polynomial in both sides of (5.5.25) by the variable t raised to the degree of the polynomial we obtain the degree-generating function identity:

$$\sum_{P \in \mathcal{M}(\mathcal{F})} t^{\text{degree}(P)} = = \prod_{i \geq 1} \frac{1}{1 - t^{\text{degree}(\phi_i)}}. \qquad (5.5.26)$$

Since the total the number of polynomials

$$P(x) = c_0 + c_1 x + c_2 x^2 + \cdots + c_{n-1} x^{n-1} + x^n \quad \text{(with } c_i \in \mathcal{F} \text{ for all } 0 \leq i < n)$$

is clearly q^n, by grouping together terms of equal degree in (5.5.26), gives the beautiful formal series identity

$$\frac{1}{1 - qt} = \sum_{n \geq 0} q^n t^n = \prod_{d \geq 1} \left(\frac{1}{1 - t^d} \right)^{I_d(\mathcal{F})}. \qquad (5.5.27)$$

Using the basic identity

$$\frac{1}{1-u} = \exp\left(\sum_{n \geq 1} \frac{u^n}{n} \right) \qquad (5.5.28)$$

for $u = qt$ and $u = t^d$ in (5.5.27) and equating the logarithms of both sides gives

$$\sum_{n \geq 1} \frac{q^n t^n}{n} = \sum_{d \geq 1} I_d(\mathcal{F}) \sum_{m \geq 1} \frac{t^{dm}}{m}$$

$$= \sum_{d \geq 1} d I_d(\mathcal{F}) \sum_{m \geq 1} \frac{t^{dm}}{dm}$$

$$= \sum_{n \geq 1} d I_d(\mathcal{F}) \sum_{n \geq 1} \frac{t^n}{n} \chi(d \preceq n)$$

$$= \sum_{n \geq 1} \frac{t^n}{n} \sum_{d \preceq n} d I_d(\mathcal{F})$$

and (5.5.24) is obtained by equating coefficients of t^n on both sides of this identity.

$\square$

Theorem 5.10 *For any integer $k \geq 1$ the polynomial $x^{p^k} - x$ factors into the product of all irreducible monic polynomials in $\mathcal{F}_p[x]$ whose degree divides k. That is, if $IM(\mathcal{F}_p)$ denotes the collection of monic polynomials in $\mathcal{F}_p[x]$ that are irreducible in $\mathcal{F}_p[x]$ we have*

$$x^{p^k} - x = \prod_{\substack{\phi \in IM(\mathcal{F}_p) \\ \text{degree}(\phi) \,\preceq\, k.}} \phi(x) . \tag{5.5.29}$$

In particular it follows from (5.5.29) that in any finite field $\mathcal{F}$ of order p^k every irreducible polynomial $\phi(x) \in \mathcal{F}_p[x]$ whose degree divides k splits into a product of linear factors.

Proof Let $\mathcal{F}$ be a finite field of order p^k and let $\Delta \subseteq \mathcal{F}$ be a set of representatives of the Frobenius cycles of $\mathcal{F}$. This given, we have the factorization

$$x^{p^k} - x = \prod_{\beta \in \Delta} \phi_\beta(x) \tag{5.5.30}$$

where $\phi_\beta(x)$ is the polynomial defined by (5.5.12). Indeed this identity is an immediate consequence of Theorem 5.6 since it simply expresses the fact that $\mathcal{F}$ breaks up into a union of disjoint Frobenius cycles. It develops that (5.5.29) and (5.5.30) are one and the same factorization. To see this let $IM_h(\mathcal{F}_p)$ denote the collection of elements of $IM(\mathcal{F}_p)$ that are of degree h and likewise let JM_h denote the collection factors $\phi_\beta(x)$ in (5.5.29) that are of degree h. Since we have shown that each $\phi_\beta(x)$ is monic and irreducible in $\mathcal{F}_p[x]$ and of degree a divisor of k we necessarily have the containment

$$JM_h \subseteq IM_h(\mathcal{F}_p) \qquad (\text{for all } h \preceq k) . \tag{5.5.31}$$

Thus to prove that the factorization in (5.5.29) and (5.5.30) are identical we need only show that we have equality in (5.5.30) for all $h \preceq k$. Now note that Proposition 5.10, for $\mathcal{F} = \mathcal{F}_p$ and $q = p$ yields the identity

$$\sum_{h \,\preceq\, k} h I_h(\mathcal{F}) = p^k , \tag{5.5.32}$$

where $I_h(\mathcal{F})$ is none other than the cardinality of $IM_h(\mathcal{F}_p)$. On the other hand if $J_h(\mathcal{F})$ denotes the cardinality of JM_h it follows from (5.5.31) that

$$J_h(\mathcal{F}) \leq I_h(\mathcal{F}) . \tag{5.5.33}$$

Now if any of these inequalities were to be strict we would have

$$\sum_{h \preceq k} h I_h(\mathcal{F}) < p^k ,$$

but that would immediately lead to a contradiction since comparing degrees of both sides of (5.5.29) yields the equality

$$\sum_{h \preceq k} h I_h(\mathcal{F}) = p^k .$$

Thus equality must hold true for all $h \preceq k$ in both (5.5.33) and (5.5.31) as desired. This proves that (5.5.29) and (5.5.30) are identical factorizations, in particular it follows that each monic irreducible polynomial in $\mathcal{F}_p[x]$ whose degree is a divisor of k splits into a product of linear factors whose roots are the elements of a Frobenius cycle of $\mathcal{F}$. This completes our proof. $\qquad\qquad\square$

The factorization in (5.5.30) has a useful by-product which is worth noting.

Proposition 5.11 *Let $\mathcal{F}$ be a field of order p^k and let Δ be a set of representatives of the Frobenius cycles of $\mathcal{F}$. Then we have the factorization*

$$Q_d(x) = \prod_{\substack{\beta \in \Delta \\ ord(\beta)=d}} \phi_\beta(x) \qquad (for\ all\ d \preceq p^k - 1) \qquad (5.5.34)$$

where again $\phi_\beta(x)$ is given by (5.5.12).

Proof The identity in (5.4.29), which holds true in all finite fields of order q, specialized to a field $\mathcal{F}$ of order p^k gives the factorization

$$x^{p^k-1} - 1 = \prod_{d \preceq p^k-1} Q_d(x) . \qquad (5.5.35)$$

On the other hand (5.5.30) gives

$$x^{p^k-1} - 1 = \prod_{\substack{\beta \in \Delta \\ \beta \neq 0}} \phi_\beta(x) = \prod_{d \preceq p^k-1} \prod_{\substack{\beta \in \Delta \\ ord(\beta)=d}} \phi_\beta(x) . \qquad (5.5.36)$$

Since the roots of $Q_d(x)$ are all the elements of order d in $\mathcal{F}$, from (5.5.35 and (5.5.36) it follows that each $\beta \in \Delta$ that is of order d must be a root of $Q_d(x)$. But since, by Proposition 5.7, p cannot be a divisor of d it follows that if β is of order d then all powers β^{p^s} are also of order d. This gives that if β is a root of $Q_d(x)$ all the elements of the Frobenius cycle of β must also be roots of $Q_d(x)$. Of course the same conclusion would be reached using the irreducibility of $\phi_\beta(x)$. Thus (5.5.34) is an immediate consequence of (5.5.35) and (5.5.36). $\qquad\square$

We are now in a position to establish one of the fundamental results in the theory of finite fields.

Theorem 5.11 *Within a field $\mathcal{F}$ of order p^k lies one and only one subfield $\mathcal{F}_h$ of order p^h for any $h \preceq k$ and it simply consists of the roots of the polynomial $x^{p^h} - x$. That is*

$$\mathcal{F}_h = \left\{ \beta \in \mathcal{F} \mid \beta^{p^h} = \beta \right\}. \tag{5.5.37}$$

Moreover, the collection $\{\mathcal{F}_h\}_{h \preceq k}$, under containment, is isomorphic to the partial order of the divisors of k. In particular all fields of order p^k are isomorphic.

Proof Let us say that an integer d *belongs to the exponent* h and write $exp(d) = h$ if h is the order of p in the integers mod d. It follows from Theorem 5.7 that an element $\beta \in \mathcal{F}$ of order d generates a subfield $\mathcal{F}_p[\beta]$ of order p^h if and only if $exp(d) = h$. We claim that for all $\gamma \in \mathcal{F}$ with $exp\big(ord(\gamma)\big) = h$ we have

$$\mathcal{F}_p[\gamma] = \mathcal{F}_h. \tag{5.5.38}$$

To show this let us pick a $d \preceq p^k - 1$ that belongs to the exponent h and any $\gamma \in \mathcal{F}$ of order d. Now if α is a primitive root of $\mathcal{F}_p[\gamma]$ then

$$\mathcal{F}_p[\gamma] = \left\{ 0, \alpha, \alpha^2, \dots, \alpha^{p^h - 1} \right\}. \tag{5.5.39}$$

In particular all elements of $\mathcal{F}_p[\gamma]$ are roots of the polynomial

$$x^{p^h} - x \; .$$

Since this polynomial cannot have more than p^h roots in $\mathcal{F}$ we see that in $\mathcal{F}$ we have

$$\beta^{p^h} - \beta = 0 \qquad \Longleftrightarrow \qquad \beta \in \mathcal{F}_p[\gamma] \; . \tag{5.5.40}$$

This proves (5.5.38). Note further that since in $\mathcal{F}_p[\gamma]$ we have the factorization

$$x^{p^h - 1} - 1 = \prod_{d' \preceq p^h - 1} \mathcal{Q}_{d'}(x) \; .$$

It follows that every element $\beta \in \mathcal{F}$, that is of order $d' \preceq p^h - 1$, must necessarily lie in $\mathcal{F}_p[\gamma]$, thus in $\mathcal{F}_h$. In particular the subfield $\mathcal{F}_p[\beta]$ it generates must entirely lie in $\mathcal{F}_h$. But if $exp(d') = h'$ then $\mathcal{F}_p[\beta] = \mathcal{F}_{h'}$. This gives

$$\mathcal{F}_{h'} \subseteq \mathcal{F}_h \qquad \Longleftrightarrow \qquad h' \preceq h \; .$$

Note further that in the particular case $h = k$ we have shown that

$$\mathcal{F}_k = \mathcal{F}_p[\beta] = \mathcal{F} \qquad \text{(for all } \beta \in \mathcal{F} \text{ such that } exp(ord(\beta)) = k) \qquad (5.5.41)$$

since the field $\mathcal{F}_p[\beta]$ is isomorphic to the quotient field $\mathcal{F}_p[x]/(\phi_\beta(x))$. From (5.5.29) we derive that if we take any polynomial $f(x) \in \mathcal{F}_p[x]$ of degree k that is irreducible in $\mathcal{F}_p[x]$ then the quotient field $\mathcal{F}_p[x]/(f(x))$ is isomorphic to $\mathcal{F}$. This proves the equality of all fields of order p^k and completes our proof. □

Here and after the unique field of order p^k will be denoted $\mathcal{F}_{p^k}$. One of the by-products of this subsection is that no matter how $\mathcal{F}_{p^k}$ is constructed it contains within itself all possible ways of constructing it. We shall call the quotient construction $\mathcal{F}_p[x]/(f(x))$ with $f(x) \in \mathcal{F}_p[x]$ irreducible of degree k a *presentation* of $\mathcal{F}_{p^k}$. But we should keep in mind that although they all yield the same field, some presentations may be more efficient than others in computer experimentations.

5.6 A Factorization Algorithm

It follows from the results in the previous subsection that every irreducible polynomial $f(x) \in \mathcal{F}_p[x]$ of degree a divisor of k splits into product of linear factors in $\mathcal{F}_{p^k}$. In this subsection we shall give an algorithm for carrying out such factorizations. Our basic tool here will be the polynomial

$$S_k(x) = x + x^p + x^{p^2} + \cdots + x^{p^{k-1}} \qquad (5.6.1)$$

for some prime p.

Proposition 5.12

(1) $S_k(x)$ maps $\mathcal{F}_{p^k}$ onto $\mathcal{F}_p$,
(2) $cS_k(x) = S_k(cx)$ for all $c \in \mathcal{F}_p$,
(3) for all $\alpha, \beta \in \mathcal{F}_{p^k}$ we have $S_k(\alpha + \beta) = S_k(\alpha) + S_k(\beta)$,
(4) If $\beta \in \mathcal{F}_{p^k}$ we have $S_k(\beta x) = 0$ for all $x \in \mathcal{F}_{p^k}$ if and only if $\beta = 0$.

Proof Note that, from Theorem 5.11 we derive that

$$S_k(\beta)^p = \beta^p + \beta^{p^2} + \cdots + \beta^{p^k} = S_k(\beta) \qquad \text{(for all } \beta \in \mathcal{F}_{p^k}) . \qquad (5.6.2)$$

This proves that $S_k(x)$ maps $\mathcal{F}_{p^k}$ into $\mathcal{F}_p$. Note further that for $c \in \mathcal{F}_p$ we have $c^{p^s} = c$ for all $s \geq 0$ thus

$$S_k(cx) = cx + c^p x^p + + \cdots + c^{p^{k-1}} x^{p^{k-1}} = cS_k(x).$$

This proves (2). Recall that in a field of characteristic p we have the identity

$$(\alpha + \beta)^{p^s} = \alpha^{p^s} + \beta^{p^s} \qquad \text{(for all } s \geq 0)$$

and this proves (3).

Next note that if $\beta \neq 0$ in $\mathcal{F}_{p^k}$ then

$$S_k(\beta x) = x\beta + x^p \beta^p + x^{p^2} \beta^{p^2} + \cdots + x^{p^{k-1}} \beta^{p^{k-1}}$$

is a polynomial of degree $p^{k-1} < p^k$ and cannot identically vanish in $\mathcal{F}_{p^k}$. This proves (4). We are left to show that $S_k(x)$ is onto. Note first that the equation $S_k(x) = 0$ cannot have more than p^{k-1} solutions in $\mathcal{F}_{p^k}$. Since $\mathcal{F}_{p^k}$ has p^k elements we can certainly find $\beta \in \mathcal{F}_{p^k}$ such that

$$S_k(\beta) \neq 0 .$$

Let then $S_k(\beta) = c_0$. Now (5.6.2) gives that $c_0 \in \mathcal{F}_p$ and from (2) we derive that

$$S_k(\beta/c_0) = 1$$

but then again from (2) it follows that

$$S_k(i\beta/c_0) = i \qquad \text{(for all } i = 0, 1, 2, \ldots, p-1) .$$

This proves that $S_k(x)$ is onto and completes our argument. $\square$

Proposition 5.13 *For any $c \in \mathcal{F}_p$ the equation*

$$S_k(x) = c \tag{5.6.3}$$

has exactly p^{k-1} solutions in $\mathcal{F}_{p^k}$.

Proof Properties (1), (2), and (3) of Proposition 5.12 yield that that S_k is a linear transformation of $\mathcal{F}_{p^k}$ onto $\mathcal{F}_p$ as vector spaces over $\mathcal{F}_p$. Since $\mathcal{F}_{p^k}$ is a k dimensional space over $\mathcal{F}_p$ and the dimension of the range of S_k is 1 it follows that the dimension of the nullspace of S_k must be $k-1$. Thus the equation

$$S_k(x) = 0$$

has exactly p^{k-1} solutions in $\mathcal{F}_{p^k}$. Now note that for any two elements $\beta_1, \beta_2 \in \mathcal{F}_{p^k}$ we have, by (3) of Proposition 5.12

$$S_k(\beta_2 - \beta_1) + S_k(\beta_1) = S_k(\beta_2) . \tag{5.6.4}$$

This shows that if $S_k(\beta_1) = c$ for some $c \in \mathcal{F}_p$ then $S_k(\beta_2) = c$ if and only if $S_k(\beta_2 - \beta_1) = 0$. Thus the solution set of (5.6.3) can be obtained by adding to a particular solution of $S_k(x) = c$ a solution of $S_k(x) = 0$. Since we have seen that $S_k(x) = 0$ has exactly p^{k-1} solutions and S_k is an onto map, it follows that the equation in (5.6.3) has exactly p^{k-1} solutions for all $c \in \mathcal{F}_p$.   □

Proposition 5.13 has the following remarkable corollary.

Theorem 5.12

$$x^{p^k} - x = \prod_{c \in \mathcal{F}_p} (S_k(x) - c) .$$ (5.6.5)

Proof Note that if

$$S_k(\beta_1) = c_1 \quad \text{and} \quad S_k(\beta_2) = c_2$$

then

$$c_1 \neq c_2 \quad \Longrightarrow \quad \beta_1 \neq \beta_2 .$$

Thus the solution sets

$$\{\beta \in \mathcal{F}_{p^k} \mid S_k(\beta) = c\}$$

are disjoint as c varies in $\mathcal{F}_p$. Since each of them has p^{k-1} elements and there are altogether p of them it follows that their union must necessarily be all of $\mathcal{F}_{p^k}$. This proves (5.6.5).   □

We are finally in a position to prove the crucial result of this subsection.

Theorem 5.13 *Let α be a primitive root of $\mathcal{F}_{p^k}$ and $f(x) \in \mathcal{F}_p[x]$ be a monic polynomial with distinct roots in $\mathcal{F}_{p^k}$. Then for each $0 \leq j \leq p^k - 1$ we have the factorization*

$$\alpha^j f(x) = \prod_{c \in \mathcal{F}_p} \gcd\left(f(x), S_k(\alpha^j x) - c\right)$$ (5.6.6)

Moreover for at least one $0 \leq j \leq p^k - 1$ this factorization is not trivial.

Proof By assumption we have

$$f(x) = (x - \beta_1)(x - \beta_2) \cdots (x - \beta_h)$$

with $\beta_1, \beta_2, \ldots, \beta_h$ distinct elements of $\mathcal{F}_{p^k}$. Now (5.6.5) with x replaced by $\alpha^j x$ gives

$$\alpha^j (x^{p^k} - x) = \prod_{c \in \mathcal{F}_p} (S_k(\alpha^j x) - c). \qquad (5.6.7)$$

Thus every β_j must be a root of the polynomial on the right. This gives the factorization in (5.6.7). Now this factorization is trivial, for a given j, only if for some $c_0 \in \mathcal{F}_p$ we have that $f(x)$ divides $S_k(\alpha^j x) - c_0$. This means that

$$S_k(\alpha^j \beta_r) = c_0 \qquad \text{(for all } r = 1, 2, \ldots, h) .$$

But then from the identity in (5.6.4) it follows that

$$S_k\big(\alpha^j (\beta_r - \beta_s)\big) = S_k(\alpha^j \beta_r) - S_k(\alpha^j \beta_s) = 0 \qquad \text{(for all } 1 \leq r < s \leq h) .$$

Now if this were to hold for all $j = 1, 2, \ldots, p^k - 1$ then from (4) of Proposition 5.12 it would follow that

$$\beta_r = \beta_s$$

contradicting the assumption that $\beta_1, \beta_2, \ldots, \beta_h$ are distinct. This contradiction proves our assertion and completes our proof.   □

Remark 5.3 Using Theorem 5.13 it is easy to put together an algorithm that completely factorizes every polynomial in $\mathcal{F}_p[x]$ with roots in $\mathcal{F}_{p^k}$.

Glossary

$(x)_a$	Lower factorial polynomial, 77	
$<_{LL}$	Young's first letter order, 1	
$A^\lambda(\sigma)$	Representation, 3	
B_n	Bell numbers, 177	
$C(T)$	Column group of T, 2	
$C(T_1, T_2)$	Signed good $T_1 \wedge T_2$, 3	
C_μ	Conjugacy class sum, 32	
$D_n(x; q)$	Denominator of convergent, 169	
D_q	q-differentiation operator, 127	
$E(w)$	End part, 141	
$E_{i,j}$	Matrix unit, 11	
F	Frobenius map, 32	
$H_n(x)$	Hermite polynomial, 179	
$I(w)$	Initial part, 141	
$INJ(\lambda)$	Injective tableau of shape λ, 1	
$J(x; c, \lambda)$	Jacobi continued fraction, 149	
$J(x; q)$	Continued fraction, 168	
$L[m, n]$	Lattice of Ferrers diagrams, 125	
$L_n(x)$	Legendre polynomial, 181	
$L_n^{(\alpha)}(x)$	Laguerre polynomial, 183	
$N_n(x; q)$	Numerator of convergent, 170	
Q_λ	Symmetric polynomial, 76	
$Q_d(t)$	dth cyclotomic polynomial, 203	
$Q_n(x)$	Orthogonal polynomials, 146	
$R(T)$	Row group of T, 2	
$R_k(x)$	Polynomial, 78	

© The Editor(s) (if applicable) and The Author(s), under exclusive license
to Springer Nature Switzerland AG 2020
A. M. Garsia, Ö. Eğecioğlu, *Lectures in Algebraic Combinatorics*,
Lecture Notes in Mathematics 2277, https://doi.org/10.1007/978-3-030-58373-6

$ST(\lambda)$	Standard tableau of shape λ, 1	
S_i^λ	ith standard tableau of shape λ, 3	
$S_k(x)$	Polynomial, 221	
S_n	Symmetric group, 1	
$S_{n,k}$	Stirling numbers, 177	
$T_1 \wedge T_2$	Diagram construction, 2	
$\alpha_n(x)$	Chebyshev polynomial 1, 164	
$\beta_n(x)$	Chebyshev polynomial 2, 164	
χ^μ	Character, 29	
ϵ	Identity permutation, 2	
γ_i^λ	Idempotent, 13	
$\gcd(A, B)$	Greatest common divisor, 188	
λ	Partition, 1	
$\lambda(T)$	Shape of tableau T, 1	
$\leq_D$	Dominance, 9	
$\mathcal{A}(S_n)$	Group algebra of S_n, 1	
$\mathcal{C}(G)$	Center of the group algebra, 12	
$\mathcal{C}(\lambda)$	Contents of λ, 79	
$\mathcal{F}$	Finite field, 187	
$\mathcal{F}(\mathcal{A}, M)$	Cartier–Foata language, 141	
$\mathcal{F}[x]$	Polynomial ring over $\mathcal{F}$, 188	
$\mathcal{F}_p$	Prime field, 192	
$\mathcal{F}_{p^k}$	Finite field, 221	
$\mathcal{M}_n[\mathbf{K}]$	Matrix algebra, 10	
$\mathcal{P}(\mathcal{A}, M)$	Pyramids of $\mathcal{L}(\mathcal{A}, M)$, 144	
$\mathcal{P}_n$	Partitions of $\{1, 2, \ldots, n\}$, 176	
$\mathcal{Z}_\rho$	Polynomial, 85	
$\mathfrak{sl}(2)$	Special linear group, 97	
μ	Möbius function, 197	
μ_n	Moment, 146	
ω_ρ^λ	Symbol, 86	
$\phi(\beta)$	Frobenius map, 209	
π_n	Operator, 82	
$\pi_{n,k}(x)$	Polynomial, 80	
σ	Permutation, 2	
$\Lambda^{=n}$	Homogeneous symmetric polynomials, 32	
Ω_p	$\{0, 1, \ldots, p - 1\}$, 191	
$\Omega_{f(x)}$	Finite field, 197	
$\Phi(n)$	Euler totient function, 199	
$\Xi_a(y)$	Polynomial, 77	

$\vert_\epsilon$	Coefficient of ϵ,	14
$d\vert n$	d divides n,	197
$d \preceq n$	d divides n,	197
$e(T)$	Group algebra element of T,	44
e_i	Elementary symmetric function,	73
f_λ	Number of standard tableaux,	82
$l(\lambda)$	Length of a partition,	79
p_λ	Power basis,	32
$p_k(x)$	Power sum symmetric function,	80
s_λ	Schur function,	82
z_μ	Symbol,	32

References

1. G. Andrews, The theory of partitions, in *Encyclopedia of Mathematical and its Applications* (Addison-Wesley, Reading, 1976)
2. G. Andrews, R.J. Baxter, A motivated proof of the Rogers–Ramanujan identities. Am. Math. Monthly **96**(5), 401–409 (1989)
3. E.R. Berlekamp, Factoring polynomials over finite fields. Bell Syst. Tech. J. **46**, 1853–1859 (1967)
4. E.R. Berlekamp, *Algebraic Coding Theory* (Aegean Park Press, Laguna Hills, 1984)
5. P. Cartier, D. Foata, in *Problèmes Combinatoires de commutation et réarrangements*. Springer Lecture Notes in Mathematical, vol. 85 (1969)
6. P. Diaconis, C. Greene, Applications of Murphy's elements. Technical Report no. 335, September 1989 (Department of Statistics, Stanford University, Stanford, 1989)
7. J. Favard, Sur les polynomes de Tchebicheff. C. R. Acad. Sci. Paris **200**, 2052–2053 (1935)
8. P. Flajolet, Combinatorial aspects of Continued Fractions. Discrete Math. **32**, 125–161 (1980)
9. J. Françon, G. Viennot, Permutations selon les pics, creux, double montées, double descentes. Discrete Math. **28**, 21–35 (1979)
10. L. Gaal, *Classical Galois Theory with Examples* (Markham Publishing Company, Chicago, 1971)
11. A.M. Garsia, S.C. Milne, A Rogers–Ramanujan Bijection. J. Comb. Theory Ser A **31**, 289–329 (1981)
12. R. Goodman, N.R. Wallach, Symmetry, representations, and invariants, in *Graduate Texts in Mathematics*, vol. 255 (2009)
13. A. Goupil, D. Poulalhon, G. Schaeffer, Central characters and conjugacy classes of the symmetric group, in *Proceedings of FPSAC'00, Moscow*, pp. 238–249 (Springer, Berlin, 2000)
14. A. Jucys, Symmetric polynomials and the center of the symmetric group ring. Rep. Math. Phys. **5**(1), 107–112 (1974)
15. R. Lidl, H. Niederreiter, *Introduction To Finite Fields and their Applications* (Cambridge University, Cambridge, 1966)
16. I.G. Macdonald, *Symmetric Functions and Hall Polynomials*, 2nd edn. (Oxford Science, Oxford, 1995)
17. G. Murphy, A new construction of Young's seminormal representation of the symmetric group. J. Algebra **69**, 287–291 (1981)
18. R.A. Proctor, Representations of $\mathfrak{sl}(2, \mathbb{C})$ on Posets and the Sperner Property. SIAM. J. Algebraic Discrete Methods **3**(2), 275–280 (1982)

A. M. Garsia, Ö. Eğecioğlu, *Lectures in Algebraic Combinatorics*, Lecture Notes in Mathematics 2277, https://doi.org/10.1007/978-3-030-58373-6

19. R.A. Proctor, Solution of two difficult combinatorial problems with Linear Algebra. Am. Math. Monthly **89**(10), 721–734 (1982)
20. D.E. Rutherford, *Substitutional Analysis* (Edinburgh University, Edinburgh, 1948)
21. I. Schur, Ein Betrag zur additiven Zahlentheorie und zur Theorie der Kettenbruche, in *S-B Preuss. Akademie Wissenschaftliche Physics Mathematical Kl* (1917), pp. 302–321
22. E. Sperner, Ein Satz über Untermengen einer endlichen Menge. Math. Zeitschrift **27**, 544–548 (1928)
23. R.P. Stanley, Weyl Groups, the Hard Lefschetz Theorem, and the Sperner Property. SIAM. J. Algebraic Discrete Methods **1**(2), 168–184 (1980)
24. R.M. Thrall, Young's seminormal representation of the symmetric group. Duke Math. J. **8**, 611–624 (1941)
25. G. Viennot, in *Une Théorie Combinatoire des Polynomes Orthogonaux Generaux*. Lecture notes UQAM Montréal (1983)
26. G. Viennot, in *Heaps of Pieces*. Lecture Notes University of Bordeaux # I-8614. 351 cours de la Libération 33405 Talence, France. See also Proceedings of the Montréal Colloquium In Enumerative Combinatorics, UQAM Montréal (1985)
27. J. von Neumann, Approximative properties of matrices of high finite order. Portugal Math. **3**, 1–62 (1942). Collected Works IV, Pergamon Press (1962), p. 271
28. The Collected Papers of A. Young, *Mathematical Expositions V.* , vol. #21, pp. 1873–1940 (University of Toronto, Toronto, 1977)
29. A. Young, On quantitative substitutional analysis (sixth paper), in *The Collected Papers of Alfred Young, 1873–1940*. Mathematical Expositions, vol. 21, pp. 432–466 (University of Toronto Press, Toronto, 1901)

LECTURE NOTES IN MATHEMATICS

Editors in Chief: J.-M. Morel, B. Teissier;

Editorial Policy

1. Lecture Notes aim to report new developments in all areas of mathematics and their applications – quickly, informally and at a high level. Mathematical texts analysing new developments in modelling and numerical simulation are welcome.

    Manuscripts should be reasonably self-contained and rounded off. Thus they may, and often will, present not only results of the author but also related work by other people. They may be based on specialised lecture courses. Furthermore, the manuscripts should provide sufficient motivation, examples and applications. This clearly distinguishes Lecture Notes from journal articles or technical reports which normally are very concise. Articles intended for a journal but too long to be accepted by most journals, usually do not have this "lecture notes" character. For similar reasons it is unusual for doctoral theses to be accepted for the Lecture Notes series, though habilitation theses may be appropriate.

2. Besides monographs, multi-author manuscripts resulting from SUMMER SCHOOLS or similar INTENSIVE COURSES are welcome, provided their objective was held to present an active mathematical topic to an audience at the beginning or intermediate graduate level (a list of participants should be provided).

    The resulting manuscript should not be just a collection of course notes, but should require advance planning and coordination among the main lecturers. The subject matter should dictate the structure of the book. This structure should be motivated and explained in a scientific introduction, and the notation, references, index and formulation of results should be, if possible, unified by the editors. Each contribution should have an abstract and an introduction referring to the other contributions. In other words, more preparatory work must go into a multi-authored volume than simply assembling a disparate collection of papers, communicated at the event.

3. Manuscripts should be submitted either online at www.editorialmanager.com/lnm to Springer's mathematics editorial in Heidelberg, or electronically to one of the series editors. Authors should be aware that incomplete or insufficiently close-to-final manuscripts almost always result in longer refereeing times and nevertheless unclear referees' recommendations, making further refereeing of a final draft necessary. The strict minimum amount of material that will be considered should include a detailed outline describing the planned contents of each chapter, a bibliography and several sample chapters. Parallel submission of a manuscript to another publisher while under consideration for LNM is not acceptable and can lead to rejection.

4. In general, **monographs** will be sent out to at least 2 external referees for evaluation.

    A final decision to publish can be made only on the basis of the complete manuscript, however a refereeing process leading to a preliminary decision can be based on a pre-final or incomplete manuscript.

    Volume Editors of **multi-author works** are expected to arrange for the refereeing, to the usual scientific standards, of the individual contributions. If the resulting reports can be forwarded to the LNM Editorial Board, this is very helpful. If no reports are forwarded or if other questions remain unclear in respect of homogeneity etc, the series editors may wish to consult external referees for an overall evaluation of the volume.

5. Manuscripts should in general be submitted in English. Final manuscripts should contain at least 100 pages of mathematical text and should always include

 – a table of contents;
 – an informative introduction, with adequate motivation and perhaps some historical remarks: it should be accessible to a reader not intimately familiar with the topic treated;
 – a subject index: as a rule this is genuinely helpful for the reader.
 – For evaluation purposes, manuscripts should be submitted as pdf files.

6. Careful preparation of the manuscripts will help keep production time short besides ensuring satisfactory appearance of the finished book in print and online. After acceptance of the manuscript authors will be asked to prepare the final LaTeX source files (see LaTeX templates online: https://www.springer.com/gb/authors-editors/book-authors-editors/manuscriptpreparation/5636) plus the corresponding pdf- or zipped ps-file. The LaTeX source files are essential for producing the full-text online version of the book, see http://link.springer.com/bookseries/304 for the existing online volumes of LNM). The technical production of a Lecture Notes volume takes approximately 12 weeks. Additional instructions, if necessary, are available on request from lnm@springer.com.

7. Authors receive a total of 30 free copies of their volume and free access to their book on SpringerLink, but no royalties. They are entitled to a discount of 33.3 % on the price of Springer books purchased for their personal use, if ordering directly from Springer.

8. Commitment to publish is made by a *Publishing Agreement*; contributing authors of multiauthor books are requested to sign a *Consent to Publish form*. Springer-Verlag registers the copyright for each volume. Authors are free to reuse material contained in their LNM volumes in later publications: a brief written (or e-mail) request for formal permission is sufficient.

Addresses:
Professor Jean-Michel Morel, CMLA, École Normale Supérieure de Cachan, France
E-mail: moreljeanmichel@gmail.com

Professor Bernard Teissier, Equipe Géométrie et Dynamique,
Institut de Mathématiques de Jussieu – Paris Rive Gauche, Paris, France
E-mail: bernard.teissier@imj-prg.fr

Springer: Ute McCrory, Mathematics, Heidelberg, Germany,
E-mail: lnm@springer.com

Printed in the United States
By Bookmasters